T0383924

Explanatory Pluralism

Explaining phenomena is one of the main activities in which scientists engage. This book proposes a new philosophical theory of scientific explanation by developing and defending the position of explanatory pluralism with the help of the notion of "explanatory games". Mantzavinos provides a descriptive account of the explanatory activity of scientists in different domains and shows how they differ from commonsensical explanations offered in everyday life by ordinary people and also from explanations offered in religious contexts. He also shows how an evaluation and a critical appraisal of explanations put forward in different social arenas can take place on the basis of different values. *Explanatory Pluralism* provides solutions to all important descriptive and normative problems of the philosophical theory of explanation as illustrated in sophisticated case studies from economics and medicine but also from mythology and religion.

C. MANTZAVINOS is Professor of Philosophy of the Social Sciences at the University of Athens. He has previously taught at Witten/ Herdecke, Freiburg, Bayreuth and Stanford and was a senior research fellow at the Max Planck Institute for Research on Collective Goods, Bonn. He has held visiting appointments at Harvard and the Maison des Sciences de l'Homme, Paris. He is the author of *Wettbewerbstheorie* (1994), *Individuals, Institutions, and Markets* (Cambridge University Press, 2001) and *Naturalistic Hermeneutics* (Cambridge University Press, 2005) and the editor of *Philosophy of the Social Sciences* (Cambridge University Press, 2009).

Explanatory Pluralism

C. MANTZAVINOS

University of Athens

Shaftesbury Road, Cambridge CB2 8EA, United Kingdom

One Liberty Plaza, 20th Floor, New York, NY 10006, USA

477 Williamstown Road, Port Melbourne, VIC 3207, Australia

314–321, 3rd Floor, Plot 3, Splendor Forum, Jasola District Centre, New Delhi – 110025, India

103 Penang Road, #05–06/07, Visioncrest Commercial, Singapore 238467

Cambridge University Press is part of Cambridge University Press & Assessment, a department of the University of Cambridge.

We share the University's mission to contribute to society through the pursuit of education, learning and research at the highest international levels of excellence.

www.cambridge.org
Information on this title: www.cambridge.org/9781107128514

© C. Mantzavinos 2016

This publication is in copyright. Subject to statutory exception and to the provisions of relevant collective licensing agreements, no reproduction of any part may take place without the written permission of Cambridge University Press & Assessment.

First published 2016

A catalogue record for this publication is available from the British Library

Library of Congress Cataloging-in-Publication data
Mantzavinos, Chrysostomos, author.
Explanatory pluralism / C. Mantzavinos.
New York : Cambridge University Press, 2016. | Includes bibliographical references and index.
LCCN 2016006869 | ISBN 9781107128514 (hardback)
LCSH: Explanation. | Social sciences – Philosophy. | Science – Philosophy.
LCC BD237 .M36 2016 | DDC 121/.6–dc23
LC record available at http://lccn.loc.gov/2016006869

ISBN   978-1-107-12851-4   Hardback

Cambridge University Press & Assessment has no responsibility for the persistence or accuracy of URLs for external or third-party internet websites referred to in this publication and does not guarantee that any content on such websites is, or will remain, accurate or appropriate.

Στήν μητέρα μου Νίκη

πού μ' ἔμαθε ν' ἀγαπῶ τά γράμματα

Contents

Preface and Acknowledgments

The *Alte Burse* is a wonderful dusky pink building with a symmetrical facade in Tübingen. Built in 1480, it is the domicile of the Philosophy Department. It borders the *Tübinger Stift* in the west where Hegel, Schelling and Hölderlin passed their years as students and the *Hölderlinturm* in the east where Hölderlin spent the last years of his life. The two separate entrances to the *Alte Burse*, which can be reached through two separate staircases built in opposite directions to each other, show how a lively philosophical dispute can leave its track on a building. This architectonic peculiarity goes back to the early modern *Universalienstreit*, which led, in Tübingen, to the erection of a wall in the building that separated the *Alte Burse* in two halves and so created a separate entrance respectively for the two "ways", the Via antiqua and the Via moderna, left for the "Nominalists" and right for the "Realists". The wall was removed only during the Reformation when conflicts of a more rebellious character broke out. But the two entrances remained, a memorial to the irreconcilable controversy between two competing approaches to what constitutes scientific knowledge, a controversy which has taken new turns and manifestations during the centuries and still calls for arbitration today.

Hörsaal X overlooking the Neckar River is the focal point of the intellectual activity in the *Alte Burse* today. It was in this room that I attended in the *Sommersemester* 1991 the Hauptseminar on Popper's "Logik der Forschung" offered by Professor Keuth which also included the analysis of Hempel's "Aspects of Scientific Explanation". The discussions on scientific explanation in this room, overlooking the Platanenallee on the Neckarinsel, have had a lasting impression on me. Although I have turned my attention to other problems over the years, I have always retained a keen interest in explanation as the core theoretical activity in science. About a decade ago, I set out to develop my own theory of explanation which has taken preliminary shape in my essay entitled "Explanatory Games", published in the *Journal of Philosophy* in 2013. The present book contains a more complete treatment of my proposal on how to think about the core theoretical activity, explanation.

I cannot pretend that my endeavour has been a successful one, if judged according to what I originally set out to accomplish. The contrary is the case. What I set out to accomplish was to develop my own model of explanation which would ideally share the positive features of the Popper-Hempel view: rigour, simplicity and universal scope. But, through the years, it has become increasingly clear to me that the development of such a unitary model of explanation would be yet another philosophical fiction vis-à-vis the actual explanatory activities in which scientists engage. And that in the end, it would share the fate of all such philosophical fictions: to be – rightly – ignored. I gradually became aware, thus, that the main task to be accomplished was to develop a general framework capable of accounting for the plurality and variety of explanatory activities in science and in everyday life. The philosophical theory that I offer in this book consists of my proposal of how *this* task can be best accomplished.

Over the years many people whom I would like to thank have influenced my thinking on the problems that I deal with in the book. The original influence has come from Herbert Keuth and Hans Albert in exchanges in Tübingen and Heidelberg. I have received decisive impulses from Philip Kitcher in many discussions that we have had in Germany, the United States and in Greece. I am very grateful for his kindness and encouragement, which have been invaluable during all the stages of the development of the book. I owe special thanks to Sandra Mitchell – my acquaintance with her integrative pluralism through her writings and through our personal exchanges has been extremely helpful to me. I am particularly grateful to Philip Pettit for many penetrating comments and for his generous support of the project. I have learnt a lot from the comments on the different versions of the typescript and from discussions from Theo Arabatzis; the combination of his perspicacity and personal warmth are unique.

I have greatly benefited from conversations with Jim Woodward at different occasions in Heidelberg, Stirling and Pittsburgh for which I am grateful. And I owe special thanks to Michael Strevens for written comments and for very useful discussions in Witten and in New York.

I have been very fortunate to be a faculty member at the Department of Philosophy and History of Science at the University of Athens while doing all the writing. Petros Gemtos and Tassos Bougas have been my main interlocutors for many years in Athens long before I joined the Department and I would like to thank them for all that I have learnt from them. For very helpful discussions and comments I want to particularly thank my colleagues Dionysios Anapolitanos, Jean Christianidis, Costas Dimitracopoulos, Kostas Gavroglou, Aris Hatzis, Antony Hatzimoysis, Katerina Ierodiakonou, Pavlos Kalligas, Vassilios Karakostas, Vasso Kindi, Eleni Manolakaki, Drakoulis Nikolinakos, Stathis Psillos, Yannis Stephanou, Stelios Virvidakis and Spyros Vliamos. I am especially grateful to Stella Vosniadou for her written comments on Chapter 8 of the book.

I owe particular thanks to Dan Garber for useful interactions in Athens and when I visited Princeton. I have also had very profitable discussions with Charles Beitz, Gilbert Harman, Frank Jackson, Alexander Nehamas, Michael Smith and Christian Wildberg while in Princeton. I received useful advice from Hasok Chang and Peter Godfrey-Smith for which I am grateful. Special thanks go to Raine Daston and Gerd Gigerenzer for exchanges on rationality in science over many years. I have had the opportunity to discuss my ideas on explanation repeatedly with Pierre Demeulenaere and Gianluca Manzo when visiting GEMASS at the Maison des Sciences de l'Homme in Paris, and I would like to thank them for their comments and their goodwill.

I have discussed the ideas of the book in a few seminars in Witten/Herdecke, and I would like to especially thank Diego Rios and Jens Harbecke for their input and contributions, as well as my graduate students there, Pablo Abitbol, Catherine Herfeld, Anna Janus and Carl-David Mildenberger. Special thanks to my previous secretary Gabriela Koerber for her support over many years.

For insights and suggestions offered on different occasions I would like particularly to thank Max Albert, Jim Alt, Michael Baurmann, Paul Boghossian, Jim Bogen, Luc Bovens, Michael Bratman, Martin Carrier, Andrés Casas-Casas, Jules Coleman, Tim Crane, Wolfgang Detel, John Dupré, Pascal Engel, Uljana Feest, Volker Gadenne, Stephan Hartmann, Carsten Held, Carsten Herrmann-Pillath, Paul Hoyningen-Huene, Andreas Hüttemann, Stathis Kalyvas, Matthias Kettner, Jack Knight, Maria Kronfeldnder, Meinhard Kuhlmann, Dan Little, Simon Lohse, Steven Lukes, Christoph Lütge, Holger Lyre, Peter Machamer, Uskali Mäki, Jason McKenzie Alexander, Phil Mitsis, John Norton, David Papineau, Thomas Reydon, David-Hillel Ruben, Thomas Sattig, André Schmidt, John Searle, Ian Shapiro, Ernie Sosa, Thomas Sturm, Marcel Weber, Torsten Wilholt and Joachim Zweynert.

Special thanks are due to Petros Bouras-Vallianatos for his useful comments on the case study from medicine contained in Chapter 6, Section 6.3 of the book.

I am also grateful to the participants of my graduate seminar on explanation in Athens for many helpful comments, especially Panos Loukas, Zaharias Flouris and Ilias Theodorakopoulos.

I am indebted to my editor at Cambridge University Press, John Haslam for his goodwill, his decisive support of the project and his skilful guidance through all the stages until the publication of the book. I would also like to thank Beatrice Rehl and Hilary Gaskin for their significant backing of the project. I owe a special thanks to Carrie Parkinson and Ian McIver for their help at the final production stages of the book. I would also like to thank Sri Hari Kumar Sugumaran for the assistance during the copy-editing process.

Darrell Arnold, as always, was the first to provide comments on the chapters of the book and has helped me remove many linguistic and stylistic errors. I am very grateful to him for his help over the years.

Talks based upon the material of this book were presented at the 14th International Congress of Logic, Methodology and Philosophy of Science in July 2011 in Nancy; at the University of Bielefeld in November 2011; at the University of Athens in March 2012; at the Technical University of Munich in June 2012; at the 86th Joint Session of the Aristotelian Society and the Mind Association in Stirling in July 2012; at the 8th Congress of the Gesellschaft für Analytische Philosophie (GAP) in Constance in September 2012; at the 1st Congress of the Gesellschaft für Wissenschaftsphilosophie (GWP) in Hanover in March 2013; at the 23rd World Congress of Philosophy in Athens in August 2013; at the Center for Philosophy of Science in Pittsburgh in November 2013; at the Universidad Tecnológica de Bolívar in Cartagena in February 2015; at the 89th Joint Session of the Aristotelian Society and the Mind Association in Warwick in July 2015; and the Munich Center for Mathematical Philosophy in October 2015. I thank the participants of these events for their comments and suggestions.

I would like to thank the editors of the *Journal of Philosophy* for the permission to use material from the essay "Explanatory Games" published in vol. CX, November 2013, pp. 606–632, and Bardwell Press, Oxford, for the permission to include parts of the essay "The Plurality of Explanatory Games" as Chapter 7 of the book, which has been originally published in *Theories and Social Mechanisms. Essays in Honour of Mohamed Charkaoui*, edited by Gianluca Manzo, 2015, pp. 325–335. I would also like to acknowledge the permission to reproduce as Figure 4 the "Circulatory System as Conceived in Ancient Times" from the *US National Library of Medicine*, as Figure 5 the "Andreas Vesalius: Tabulae Anatomicae Sex, Broadside 3, 1538" from the *Special Collections Department of the Glasgow University Library*, as Figure 6 the "Anatomical

Drawing of the Heart by Leonardo Da Vinci" from the *Royal Collection Trust*, as Figure 7 the image of "Plate II from Hieronymus Fabricius ab Aquapendente, De Venarum Ostiolis, 1603" from *TheMitralValve.org*, as Figure 8 the image of "Figures 1–2 from Folding Leaf in Harvey's Exercitatio Anatomica de Motu Cordis et Sanguinis in Animalibus from the 1628 edition, Frankfurt: Sumptibus Guilielmi Fitzeri" also from *TheMitralValve.org*, and as Figure 9 the image of "Malpighi Marcello, De Pulmonibus (1661/1929, p. 11) showing the Lungs of a Frog with a Cross Sectional Microscopic View" from the *Wellcome Trust*.

My greatest debt is to my family, especially to my wife Georgia for her love and affection.

Athens, 29 December 2015

I Introduction

A philosophical theory of explanation must solve many problems. It must provide a descriptive account of the explanatory activity of scientists in the different domains of science. It must show how this activity differs from the provision of commonsensical explanations permanently offered in everyday life by ordinary people and how the explanatory practices evolve over time in different social arenas. It must also account for the fact that explanatory activities very often, if not always, take place under conditions of a cognitive division of labour in which different participants to the explanatory enterprise undertake different tasks and assume different roles. In a nutshell, a philosophical theory of explanation must be able to provide an adequate descriptive account of the different facets of the explanatory activity, a task which is much harder than it prima facie appears.

Besides, it must also provide standards for judging the quality of the outcomes of the explanatory activity. Some explanations are better in certain respects than others, and a set of norms is necessary for providing judgements of their quality. Some explanations provided in everyday life and in science can be more accurate, simpler or closer to the truth than alternatives that are on offer. They might be able to provide better understanding of the subject matter that they are supposed to deal with, and they might be much easier to use when one wants to intervene in the world on the basis of them. In short, normative rules of adjudicating between the explanations offered in different social contexts are needed and must be worked out and debated critically.

These diverse aims that a philosophical theory of explanation must accomplish are of both a *descriptive* and a *normative* nature. Stated differently, a philosophical theory of explanation is to provide

1

solutions to a series of problems, both descriptive and normative, using not only philosophical resources but whatever resources are also available from other disciplines. The aim of this book is to establish the claim that this can be best done if one theorizes in terms of explanatory games rather than focusing on the explication of the concept of explanation. This is one way to defend the position of explanatory pluralism, and possibly the most successful one.

2   The Wrong Question: What Is an Explanation?

Though most of the thoughts and arguments in currency in the modern theory of explanation can be found one way or another in the works of some past philosopher from Aristotle to John Stuart Mill, the great advancements in logic and the explosion of technical sophistication of the philosophers of the twentieth century have led to a more systematic treatment of scientific explanation, giving birth to a whole sub-discipline in the philosophy of science dealing with this issue.[1] Since the classic contribution of Hempel and Oppenheim (1948), the modern theory of explanation has largely come to reflect the virtues and vices of the analytic tradition: precise formulations and carefully exposed arguments on the one hand, but also a passionate insistence on logical aspects at the cost of more substantial aspects on the other. The primary question that the philosophical theory of explanation has tried to answer has been "what an explanation consists in". As an answer to this question, Hempel has famously maintained that to explain a singular event is to show how this event can be expected to happen if one takes into account the laws that govern its occurrence together with its initial conditions. For Hempel, an explanation is a valid deductive argument whose premises include law-like statements and initial conditions and whose conclusion states that the event to be explained did occur. An explanation amounts to a statement of the nomic expectability of the explanandum event, and the concept of explanation is according to Hempel primarily epistemic.

---

[1]   For a competent and precise shorter review starting with Aristotle and including the main models of explanation prevailing in the present discussion, see Psillos (2007). See also the monographs by Salmon (1990), Psillos (2002), Ruben (2012) and Weber *et al.* (2013).

One set of reactions to the so-called received view of scientific explanation[2] has centred around the provision of counterexamples to Hempel's model, ranging from shadows explaining the height of poles[3] to magicians hexing salt[4] and much more. These counterexamples were originally meant to highlight different difficulties of this account having to do with its specific features, that is, that it did not invoke any notion of causality, that it did not problematize sufficiently the pragmatic aspect of explanations and so on. However, all the drawbacks of this account are really symptoms of a more general problem: Hempel's covering-law model of explanation was ultimately designed as a unitary model that was supposed to account for *all and every* kind of explanation provided in the different domains of science.

The more constructive critics of Hempel offered alternative models of explanation showing how they could better account for the cases in which Hempel's model failed. Just to name the most influential ones: the *causal mechanistic model*, which claims that an explanation consists in the identification of mechanisms understood as entities and activities organized such that they are productive of regular changes from start to termination conditions;[5] the *unification model*, which

2   See Salmon (1990, p. 8), who has coined this term.

3   In the literature the vertical flagpole example is usually referred to, but in his seminal paper Bromberger refers to the height of a telephone post to which a taut wire is connected – exemplifying, of course, the same point. See Bromberger (1966, p. 105).

4   See Kybourg (1965) and Salmon (1970).

5   The causal/mechanistic approach to scientific explanation was born mainly as an attempt to repair the two most serious problems of the received view, that is, the problem of causal asymmetries (associated with the famous flagpole counterexample) and the problem of relevance (associated with the famous example of hexing the table salt). Besides Railton (1978, 1981) and Humphreys (1981, 1989), it is Wesley Salmon who has most prominently argued in favour of bringing "cause" back into "because". The straightforward way to remedy the main problems of the Hempel–Oppenheim model is supposed to consist in integrating a theory of causality into the theory of explanation or, in other words, in providing scientific explanations by identifying causes of events and/or processes. Since this approach tries to take account of the explanatory practices in science (mainly physics), it does not only aim at derivations of low-level laws and generalizations from higher-level theories but also at elucidating the mechanisms at work. To explain is, thus, to expose the internal workings, to lay bare the hidden mechanisms, to open the black boxes that nature presents to us. This view makes explanatory knowledge into knowledge of the hidden mechanisms by which nature works. Salmon (1984) has tried to specify the notion of mechanism by

claims that explanations are deductive arguments that provide understanding by fitting the particular facts and events within a general theoretical framework;[6] the *pragmatic account* of explanation, which

pointing to causal processes: according to his theory, those processes (and only those)
are causal that are capable of transmitting a mark. Besides, he has endorsed an ontic
conception of explanation rejecting the epistemic and modal conceptions. (For
a criticism of the ontic conception see Wright (2015).)

The first decade of the new millennium has seen an explosion of work in this
direction. On the one hand, the seminal paper of Machamer *et al.* (2000) has provoked
further the "Thinking About Mechanisms". On the other hand, further work on
causality has been produced, and the accounts of causality have reached a higher
level of technical sophistication than any of the accounts in the past (Cartwright
2007). Mechanisms as "entities and activities organized such that they are productive
of regular changes from start or set-up to finish or termination conditions" (Machamer
et al. 2000, p. 3) should be sought in order to be able to explain how a phenomenon
comes about or how some significant process works – this is the main message of the
mechanistic approach to explanation which comes, of course, in different variations
(Glennan 2002, Colombo *et al.* 2015). The search for mechanisms goes hand in hand
with three claims: (1) explanations should provide causes (or reasons); (2) explanations
should make phenomena intelligible; (3) explanations should exhibit the continuity
among the explaining parts (Machamer 2009). Whereas Salmon's causal/mechanical
approach was mainly inspired by physics, and his mark-transmission theory of causal
processes was tailored to physics (and later in Salmon (1998) and also in Dowe (2000)
the conserved quantity-transmission), the modified mechanistic approach has
extended its reach to the life sciences, the cognitive sciences and the social sciences.
Defenders of this approach in the life sciences claim that mechanistic explanations
differ from more traditional, nomological explanations because (a) they are not limited
to linguistic representations and logical inference but employ frequently diagrams to
characterize mechanisms and simulations to reason about them; (b) the fact that
mechanisms involve organized systems of component parts and operations provides
direction to both the discovery and testing of mechanistic explanations; and (c) models
of mechanisms are developed for specific exemplars and are not represented in terms of
universally quantified statements (Wimsatt 1976, p. 671, Bechtel and Abrahamsen
2005, Bechtel 2006, 2011, Darden 2006, Bogen and Machamer 2011). In the cognitive
neurosciences, the mechanistic approach points to the fact that explanations in neuroscience describing mechanisms are multilevel and integrate multiple fields (Craver
2007, Bechtel 2008, Harbecke 2010). Finally, in the social sciences, a great number of
both philosophers and practising scientists hold the view that social scientific explanations require the discovery of the underlying causal mechanisms that give rise to the
outcomes of interest (Hedström and Swedberg 1996, Schmid 2006, Demeulenaere
2011). The search for causal mechanisms is often combined not only with the position
of methodological individualism (Elster 2007) but more recently also with the position
of methodological localism (Little 2009, Knight 2009). It has also been argued that
narrative explanations in historical science are descriptions of epiphenomenal mechanisms (Glennan (2010) and a critical discussion by Currie (2014)).

6   The unification thesis, whose chief proponents are Friedman (1974) and Kitcher (1981,
1985, 1989), holds that scientific understanding increases as we decrease the number of
independent assumptions that are required to explain what goes on in the world.

claims that explanation is not a relationship like that of description, that is, a relationship between theory and fact, but rather a three-term relationship, that is, between theory, fact and context;[7] the *manipulationist account* of explanation, which claims, relying on invariant generalizations rather than covering laws, that an explanation primarily answers a "what-if-things-had-been-different question", that is, that an explanation primarily enables us to see what sort of difference it would have made for the explanandum if the factors cited in the explanans had been different in various possible ways;[8] and the *kairetic model*, which claims

It seeks laws and principles of high generality with the aim of constructing a coherent world picture and fitting particular facts within this framework. Besides, it is not committed to the world picture being deterministic since it is perfectly compatible with the position that basic laws can be irreducibly statistic. The thrust of the central argument of this approach is nicely summarized in the following quote from the classic paper of Michael Friedman (1974, p. 15): "I claim that this is the crucial property of scientific theories we are looking for; this is the essence of scientific explanation – science increases our understanding of the world by reducing the total number of independent phenomena that we have to accept as ultimate or given. A world with fewer independent phenomena is, other things equal, more comprehensible than with more." The unification approach, being, of course, different from the received view of Hempel–Oppenheim, still remains somehow close to it. This is mainly in virtue of its insistence on deductivism, as is the case in Kitcher's approach which focuses on the scarcity of patterns of derivation. Unification, according to Kitcher, is reached by deriving descriptions of many types of phenomena using one or a few argument patterns over and over again respecting certain constraints, stringency being the most important one. Some new and important work on the unification approach has been produced which stresses different dimensions of unification (e.g., Schurz and Lambert 1994, Schurz 1999, Bartelborth 2002, Bartelborth 2007, ch. 6, Colombo and Hartmann 2015, Nathan 2015, Petkov 2015, and the criticism of Gijsbers (2007) and for a review Psillos (2002)) and it is characteristic for its relevance that any new theoretical endeavour on scientific explanation feels obliged to take a position vis-à-vis this approach.

7   According to van Fraassen (1980), one of the most prominent defenders of the pragmatic account of explanation, the discussion of explanation went wrong at the very beginning because explanation was conceived of as a relationship like that of a description, that is, a relationship between theory and fact. However, it is really a three-term relationship, that is, between theory, fact and context. Both van Fraassen and Achinstein (1983) claim that explanations are answers to why-questions. Why-questions are essentially contrastive: the question "Why P?" is elliptical for "Why P rather than P′, P″ …?" The same words can thus pose different contrastive why-questions. This account of explanation aims at completing the syntactic and semantic aspects of any exposition of scientific explanation by highlighting the pragmatic side of it. See also Faye (2012, ch. 3) and Faye (2014, pp. 183ff.).

8   The manipulationist approach of Jim Woodward is designed as an alternative to the common view that explanation involves subsumption under laws. According to Woodward (2000, 2003, 2014), whether or not a generalization can be used to explain

that explanation is a matter of finding, by way of using a kairetic criterion, which of the causal influences on a phenomenon are relevant to its occurrence, demanding more specifically that the explanation is not missing parts and that every aspect of the causal story represented by the explanatory model makes a difference to the causal production of the explanandum.[9]

All these models of explanation – even though their designers intended them to be alternatives to Hempel's model – still remain in the same research tradition of producing unitary models in order to capture what is supposedly the main aim of (theoretical) science, explanation.[10] If not explicitly, then at least implicitly, the main philosophical project of the philosophers working in this tradition

[9] has to do with whether it is invariant rather than with whether it is lawful. A generalization is invariant if it is stable or robust, in the sense that it would continue to hold under a relevant class of changes. For example, a generalization can be invariant even if it has exceptions or holds only over a limited spatio-temporal interval. A relationship among some variables (or magnitudes) X and Y is said to be causal if, were one to intervene to change the value of X appropriately, the relationship between X and Y would not change *and* the value of Y would change. In a nutshell, an explanation for Woodward ought to be such that it can be used to answer what he calls a "what-if-things-had-been-different question", that is, the explanation must enable us to see what sort of difference it would have made for the explanandum if the factors cited in the explanans had been different in various possible ways. For a very useful recent discussion of Woodward's account see Franklin-Hall (2014).

[9] Strevens (2008) proposes the merging of the causal and the unification approach in what he calls the kairetic account of explanation – inspired by the ancient Greek *kairos*, meaning a crucial moment. His approach is clearly a causal account of explanation which claims to have some of the advantages of the unification approach: explanation consists in difference-making of one or more causal factors, and those explanations that are in principle reductionist and expressed in the vocabulary of physicalism are deemed better explanations. According to this approach, the relations of causal influence are invariably physical, so that all sciences, including the social sciences, must state causal relationships using a physical vocabulary. For a discussion of the kairetic account along with the manipulationist account of Woodward see Jansson (2014).

[10] This is also true for those philosophical accounts of explanation which have been developed in parallel or immediately after Hempel's covering-law model, as, for example, the account of Karl R. Popper as formulated in his *Logik der Forschung* (1934) and in "Naturgesetze und theoretische Systeme" (1949) and later in his "The Aim of Science" (1957), and also the account of Richard B. Braithwaite in his *Scientific Explanation. A Study of the Function of Theory, Probability and Law in Science* (1953) and of Ernest Nagel in his *The Structure of Science: Problems in the Logic of Scientific Explanation* (1961).

consists in offering an explication of the concept of explanation or, stated more neutrally, in answering the "What is an explanation?" question. The development of the precise meaning of the concept of scientific explanation occupies centre stage in all those approaches.[11] Nobody can oppose – and I do not either – the famous dictum of John Stuart Mill (1843/1974, p. 464) that "[t]he word *explanation* occurs so continually and holds so important a place in philosophy, that a little time spent in fixing the meaning of it will be profitably employed". However, this cannot be but a mission of peripheral importance to a philosophical theory of explanation for the simple reason that the outcome of this endeavour can only be a more precise concept of explanation – to be used in the discourse about the solution of the descriptive and normative problems of scientific explanatory activity making up the core of the philosophical enterprise.[12]

But even if one disagrees about the nature of the philosophical enterprise that a philosophical theory of explanation should launch and follow,[13] a cursory glance at the prevailing scientific practices shows that the unitary models of explanation on offer have only a limited range of application. It is simply a matter of fact, thus, that their resources cannot capture anything but the explanatory activities of *some* areas of (theoretical) science. The claim that each of them

[11]   And in nearly all other approaches. See, for example, Halonen and Hintikka (2005, pp. 55ff.) and the discussion of two senses of explanation – the subjectivist and the objectivist – by Bird (2005).

[12]   Surprisingly, the only text that endorses this view is Noretta Koertge's "Explanation and Its Problems" (1992, p. 86): "What strikes me as unsatisfactory about the current philosophical discussion of explanation is not its failure to match our intuitions about flagpole shadows or mayors with paresis. Rather it is the paucity of explicit *theories* of explanation – the absence of systematic philosophical generalizations in which the competing explications or models of explanation play a central role. I suggest that we reverse the order of investigation. We should begin by asking what *problems* a good theory about scientific explanation might reasonably be expected to solve. Only then can we begin to sketch such a theory. [...] I believe it is only by focusing on the philosophical problem-situation that we can transcend the forty years of explication of *explanation* [...]."

[13]   For a discussion of the level of generality of theories of explanation, see Nickel (2010).

raises, that is, that it is supposed to accommodate *all and every* scientific activity, is not tenable. I will briefly focus on the case of the social sciences, providing three examples in a very sketchy form that are intended to show that the unitary models of explanation have at best limited application.

3   A Brief Outlook on the Social Sciences

The social sciences constitute a very disparate domain of science replete with debates and all kinds of controversies, and this is not the place to even start reviewing them. I will only select three fields in order to exemplify my argument. Since the three more influential models of explanation are the causal mechanistic model, the unification model and the manipulationist model, I will show that only one of those three models is suited to each case that I will discuss. My aim is not only to show that one philosophical model of explanation fits the respective case but also that the other two do not.

3.1   NEOCLASSICAL MICROECONOMIC THEORY

The most theoretically developed field of the social sciences is probably neoclassical microeconomics: It avails of a well-developed mathematical formulation, has a great range of applications and is indeed the only piece of social scientific knowledge that is offered in a standardized way in every single economics textbook. The standard neoclassical microeconomic theory is based on the theoretical construction of utility maximization. Since the marginalist revolution in the 1870s and the pioneering works of Carl Menger (1871), William Stanley Jevons (1871) and Leon Walras (1874), a theory of price has been devised based on marginal utilities. Alfred Marshall's Principles of Economics (1890/1920) then provided a systematic account of the interplay between demand and supply on product and factor markets. Today, neoclassical microeconomic theory provides a standard axiomatization of the behaviour of households and firms in markets. The general theoretical framework that underlies this neoclassical theory of markets is the rationality

hypothesis.[1] The hypothesis of utility maximization plays the fundamental role in driving economic research, and it claims to offer
a theoretical account covering all those cases where two or more individuals exchange goods under conditions of scarcity. Neoclassical
microeconomic theory offers a global theoretical framework for both
a partial analysis of a single market and a total analysis of all markets in
an economy, using the utility maximization hypothesis and focusing
on the properties of economic equilibrium.

As Mäki (2001 and Mäki and Marchionni 2009), correctly points
out, the ideal of explanatory unification is prevalent in economics, and
more specifically in neoclassical economic theory. One of the great
merits of the theory is supposed to consist precisely in its unifying
power, that is, its ability to subsume a great range of economic phenomena under a unique descriptive and explanatory scheme. Kitcher's
(1989, p. 432) verdict seems to fit exactly to the case of neoclassical
microeconomics: "Science advances our understanding of nature by
showing us how to derive descriptions of many phenomena, using the
same patterns of derivation over and over again, and, in demonstrating
this, it teaches how to reduce the number of types of facts we have to
accept as ultimate (or brute)." The unificationist approach to scientific
explanation seems unambiguously to capture, thus, the heart of the
neoclassical economic enterprise.

More specifically, the general argument pattern that lies at the
heart of the neoclassical enterprise is one of maximization under
constraints. There is a long series of specific argument patterns: the
utility maximization of consumers, the profit maximization of firms,
the vote maximization of politicians in democratic politics, etc. More
and more specific argument patterns can be embedded into the larger
argument pattern of maximization under constraints, however, and
thus provide explanations by means of unification.

Now, neoclassical microeconomic theory is, and has always
been, under serious attack (Mantzavinos 1994). Alternative research

---

[1]   For a thorough and original discussion see Herfeld (2012).

programmes have always prevailed. Institutional Economics (Mantzavinos 2001, Mantzavinos *et al.* 2004) is the most prominent case at hand (with a few Nobel Prizes awarded to a series of leading figures working in this research programme), but also more recently Behavioral Economics. This is neither the place to proceed with the evaluation of the alternative research programmes in economics nor to present the battery of critical arguments against neoclassical economics which have been shaped and elaborated on by a series of economists and philosophers. What is important here is the following: there is an *opinio communis* among critics and defenders of neoclassical economics alike that one, if not the main, virtue of the approach is its power to unify diverse phenomena under a single descriptive and explanatory scheme.

The ideal of unification has been pushed even further by applying the neoclassical toolbox to the study of phenomena other than the traditional economic ones. James Buchanan (Buchanan and Tullock 1962, Buchanan 1975, Brennan and Buchanan 1985), Anthony Downs (1957) and Mancur Olson (1965) – just to name a few of the protagonists who introduced neoclassical economic theory into the study of politics – have given birth to a whole new discipline, Public Choice (Mueller 2003), which is a consistent step towards a unification of economic and political phenomena. Besides, the same patterns of derivation of neoclassical economics "are used over and over again" to explain such diverse social phenomena as crime and family (Becker 1976, Becker and Posner 2009), but also law and sex (Posner 2010).

It is clear that the unificationist theory of explanation can provide a better account and accommodate better the practices of the scientists working within the neoclassical economic theoretical framework than the alternative philosophical theories of explanation. Taking into consideration that neoclassical economic theory avails of a unique reputation amidst the social sciences – rightly or wrongly so is not under dispute here – the unification approach to explanation seems to be tailored to capture the reasons for its popularity.

The alternative models cannot adequately capture these practices. Let us start with the causal-mechanistic model of explanation. This requires the identification of mechanisms which will ultimately provide explanatory information. In the words of Salmon (1984, p. 260), it is the "underlying causal mechanisms [...] which hold the key to our understanding of the world". This is because "causal processes, causal interactions, and causal laws provide the mechanisms by which the world works; to understand *why* certain things happen, we need to see *how* they are produced by these mechanisms" (Salmon 1984, p. 132). However, microeconomics textbooks never pay any attention to "causal processes, causal interactions, and causal laws", in the analysis of any type of market, be it the competitive market, a monopoly or an oligopoly. The endeavour is to determine precisely the equilibrium price and quantity without paying any attention to causal interactions. The causal-mechanistic model is clearly not the fitting model here.

The same is true for the manipulationist model of Woodward. Indeed, a notorious criticism of the mainstream neoclassical theory is that it operates at a level of abstraction that makes it extremely difficult to test the theory empirically. The well-worked-out deductive arguments employed to analyse market behaviour are not amenable to an analysis of the manipulationist approach. The equations that make up neoclassical economic theory are not offered as invariant generalizations that continue to hold under some range of interventions. In fact, contrary to the model of Woodward, it is required that they be invariant under all possible changes and interventions. The decision calculi that are offered in neoclassical economic theory can definitely not be accommodated by the manipulationist model, which focuses on formulating the necessary and sufficient conditions for a generalization in order for it to be descriptive of a causal relationship, that is, that it be invariant under some appropriate set of interventions. The decision calculi of neoclassical economic theory are clearly argument patterns that can be only accommodated by the unification model of explanation.

3.2   SOCIAL MECHANISMS IN SOCIOLOGY

In sociology and political science, on the other hand, a great bulk of the work of the practising scientists is concerned with the identification of social mechanisms and the description of their working properties. Take the well-observed phenomenon in the liberal democracies of the Western world that the expansion of educational opportunity does not increase social mobility nor reduce social inequality. Raymond Boudon (1974) has provided a widely accepted explanation of the puzzle that although education does provide opportunity for individuals to improve their social position, this does not translate into change in the social structure. In every educational system, at some point in the curriculum, the individual has to make some choices about staying or not staying at school. These decisions are always taking place within a certain social context and give rise to a specific educational performance. Boudon distinguishes between primary and secondary effects of social origin. The primary effect directly influences the educational performance of the children as a joint result of class-specific primary socialization processes such as the potential of parents to actively support their child in school, the general cultural differences, etc. In the words of Boudon: "The lower the social status, the poorer the cultural background – hence the lower the school achievement" (Boudon 1974, p. 29).

The other mechanism, called secondary effect, focuses on how families from different social backgrounds evaluate higher education tracks. The prospects of success of children of working-class families in higher education tracks are estimated as rather low by their parents since the parents are mostly not familiar with this type of school themselves. A major interest of the parents themselves, which influences decisively their decision making, is status maintenance, that is, they want to ensure that their current status is maintained. A consequence of this is that their children receive sufficient praise and acceptance in their family environment even if they do not necessarily aim at the highest academic achievement. Parents in families with a higher social background, on the other hand, are motivated to provide their own children with the best possible education in order to make sure that their intergenerational status level remains relatively stable. This mechanism gives rise to parents from unequal social status producing children through a family socialization process that make use of unequal educational opportunities. The differentially educated children – even in partly meritocratic societies as in the modern Western world – often just replicate the social statuses of their parents.

In a nutshell, even though education does provide opportunity for individuals to improve their social position, this does not translate into change in the social structure. Rates of social mobility have not increased over time although the set of educational opportunities has increased, and it is not the case that societies with more expansive educational systems have higher mobility rates. "Other things being equal [...] educational growth [...] has the effect of increasing rather than decreasing social and economic equality, even in the case of an educational system that becomes more equalitarian" (Boudon 1974, p. 187).

The causal mechanistic model of explanation seems to best capture this and similar cases of explanatory practice prevalent in sociology, political science, psychology and the other social sciences. The scientific endeavour of the practising scientists consists in uncovering mechanisms in the social world, and it certainly offers a "local" rather than "global" understanding. In the example that we have been discussing, Boudon's explanation does not apply to societies outside the Western liberal democracies and does not aspire to provide insights into the causal structure of the educational systems of, say, the developing countries or those of the Middle Ages.

It is clearly not the case that one looks for global theoretical frameworks and attempts to fit the particular instances under them. There is no formalized argument pattern provided in the concrete example of Boudon, nor is there any attempt to embed a specific argument pattern into a more general one. The philosophical model of explanation as unification is not fitting in this case or in similar cases in the social sciences.

Nor is the manipulationist model. The central notion of intervention is not employed by Boudon. No regression equation is involved in his explanation, or any type of counterfactual reasoning. The same is true for a series of other mechanistic explanations provided by practising social scientists – the manipulationist model cannot accommodate them.

3.3 ECONOMETRICS

Subsequently, let us have a very brief look at a purely empirical part of social science research, econometrics. The attempt here is to estimate structural equation models. To keep the discussion simple here, I will focus on the main ingredient of structural models, that is, regression equations, which specify a functional relationship between dependent and independent variables, taking into consideration that the relationship contains a random element, the "measurement error". Probably the simplest case of a regression equation is a linear regression taking the form:

$$Y = B_1X_1 + B_2X_2 + \ldots + B_nX_n + U$$

$X_1, X_2, \ldots X_n$ are the independent variables, $B_1, B_2, \ldots B_n$ are the coefficients, Y is the dependent variable, and U the error term (which makes the model a stochastic one).

The manipulationist model of explanation best accounts for this set of practices in the social sciences. Woodward's theory offers what he calls a "natural causal interpretation of regression equations":

> [A regression equation] may be understood as claiming that each of $X_1, \ldots X_n$ are direct causes of Y [...]. These causal relationships are understood as holding for each individual in the population of interest, in the sense that for each such individual [the regression equation] characterizes the response of the value of Y possessed by

that individual to some range of interventions that change the values of $X_1, \ldots X_n$ for that individual. On this interpretation, the error term U also has a causal interpretation: it represents the combined influence of all the other causes of Y besides $X_1, \ldots X_n$ that are not explicitly represented in [the regression equation].

> *(Woodward 2003, p. 320)*

In those cases where the explanatory activity consists in establishing the truth of quantitative relationships in specific populations and under specific conditions, it is clearly the manipulationist model of explanation which is the more appropriate. However, it is characteristic for the nature of philosophical enterprise that, even if Woodward's account seems to be tailored to much of the applied work in the social sciences, it still stands in the long tradition of unitary models aiming at explicating the meaning of the concept of explanation.

> I distinguish this *interpretive* issue from issues about the conditions under which one can reliably estimate the coefficients in [a regression equation]. The latter issues are epistemological: they have to do not with what [a regression equation] *means* but with when and how one can determine the values of the coefficients in [a regression equation].
>
> *(Woodward 2003, ibid)*

The unification model clearly cannot capture the econometric practices. There are no argument patterns involved to be embedded in more general argument patterns, and there is a clear commitment to causal relationships that involves counterfactual reasoning – all aspects that are intentionally left out of the unification model.

Though the generalizations that are offered in econometrics invoke causal talk and causal inference, they never identify a causal *mechanism* proper. These generalizations constitute a sort of causal knowledge, which although not possessing the dignity of laws do depict invariances which offer satisfactory explanatory information.

Regression equations and structural models do not specify mechanisms and cannot, thus, be adequately captured by the causal mechanistic model either.

These three examples show that the unitary models of explanation cannot account for all and every explanatory activity in the social sciences. They are, if at all, only partially relevant as means of description and/or normative standards, at least in the social sciences, though a similar case can be made for other parts of science as well.[2] What does this teach us then?

---

[2]   Salmon (1998, ch. 4) made a similar point about the physical sciences, suggesting that causal models were better suited to explanations of individual events, unification models to high theory. For a discussion see de Regt (2006).

4   Towards Explanatory Pluralism

The lesson that one can draw from this brief overview is that *each* of the three main models currently on offer, the unification the mechanistic and the manipulationist, can accommodate *some* of the existing scientific practices in different social scientific domains. Though they are well-worked-out models of scientific explanation and they can definitely be applied successfully in different and distinct domains of (social) scientific knowledge, their main claim – tied to their design as unitary models of scientific explanation – cannot be upheld: none of them can successfully claim the monopoly as the one and correct account of scientific explanation. In different domains of science different kinds of scientific practices prevail; so trying to establish a monolithic philosophical theory of explanation that is supposedly good for everything cannot be an acceptable strategy. In a nutshell, the goal of a philosophical account of explanation should not be to capture *the* explanatory relation, but rather to capture the many ways in which explanations are provided in the different domains of science. The position that seems obvious and which I wish to adopt is that of an explanatory pluralism, which allows for different ideal types of explanation, that is, different exemplary accounts, intended to provide a classification of different types of explanatory activities that are offered in diverse domains of science.[1]

---

[1]   An anonymous reviewer expressed the following concern: Even if the point is granted that none of the models of explanation considered in the text can accommodate each of these examples, this does not eliminate the possibility that some other, yet to be articulated, unitary account can accommodate all of the examples. In other words, the argument presented here presumes that the set of developed unitary accounts includes the best of all the possibilities, and then concludes from their inadequacies that pluralism is preferable. This is an important point and if indeed a better unitary account will be developed in the future, the argument in the text must be extended accordingly and it must be inquired whether it is still valid.

Before elucidating the position of explanatory pluralism further in a more positive way, I will state in this chapter what this position does not amount to.

4.1   EXPLANATORY PLURALISM IS INDEPENDENT OF THE NATURE OF CAUSALITY

A core issue is whether explanations must necessarily include causes, or in other words whether any genuine explanation has to take into consideration the causes involved. The causal mechanistic approach, the manipulationist account (and also the kairetic account) take this for granted and go on to specify how exactly causality is tied to explanation. The unification approach, on the other hand, contends that the concept of causal dependence is derivative from that of explanatory dependence. It is from the "because" that we can infer the "cause" and not vice versa. What is distinctive about the unification view is, thus, that it proposes to ground causal claims in claims about explanatory dependency rather than vice versa. This approach accounts for the intuition that appeals to shadows not explaining the heights of towers because shadow heights are causally dependent on tower heights. It suggests that our view of causality in this and similar cases rather stems from an appreciation of the explanatory ordering of our beliefs. Put more simply, we can only identify something as a cause because this something provides the explanation of the phenomenon at hand – the concept of causal dependence is derivative from that of explanatory dependence. Adopting a position and making a decision in favor of the primacy of causal or explanatory dependence goes hand in hand with adopting the respective metaphysical commitments and coming to grips with Hume's legacy that causal judgements are epistemologically problematic.

This difference in the approaches seems to me to be really fundamental since it concerns the thorny metaphysical question about the nature of causality. Philosophers dealing with the problem of causation tend to appeal in their work to quite sophisticated metaphysical machinery trying to specify what a cause consists in exactly, ranging from manipulability and causal dependence to much more.[2] "The law of causality", however, as Bertrand Russell famously stated in 1912, "like much that passes muster among philosophers, is a relic of a bygone age, surviving, like the monarchy, only because it is erroneously supposed to do no harm".

Though more and more philosophers hasten to jump on the train of causality in order to reach explanation, cautiousness is necessary. Not all explanation is causal explanation; the most prominent examples of non-causal explanation include mathematical,[3] geometrical

---

[2]   For a critical review of different views of causality see Cartwright (2007, ch. 4).

[3]   Steiner's view of mathematical explanation has long been influential (1978a, p. 143): "My proposal is that an explanatory proof makes reference to a characterizing property of an entity or structure mentioned in the theorem, such that from the proof it is evident that the result depends on the property. It must be evident, that is, that if we substitute in the proof a different object of the same domain, the theorem collapses; more, we should be able to see as we vary the object how the theorem changes in response. In effect, then, explanation is not simply a relation between a proof and a theorem; rather, a relation between an array of proofs and an array of theorems, where the proofs are obtained from one another by [a process of] 'deformation' [...]."

Building on this notion of mathematical explanation, Steiner (1978b, p. 19) has worked out "[t]he difference between mathematical and physical explanations of *physical* phenomena [...] In the former, as in the latter, physical and mathematical truths operate. But only in mathematical explanation is this the case: when we remove the physics, we remain with a mathematical explanation – of a mathematical truth!"

In his illuminating discussion of mathematical explanations Lange (2014, p. 487) correctly observes that in mathematical explanations consisting of proofs, the source of explanatory priority cannot be causal or nomological or temporal priority, but rather at least part of its source seems to be that axioms explain theorems, not vice versa. Besides the priority of axioms over theorems in mathematics, Lange (2014, p. 525) "also emphasized mathematical explanations that operate in connection with 'problems,' each of which is characterized by a 'setup' and a 'result'. This structure of setup and result adds an asymmetry that enables mathematical explanation to get started by allowing why questions to be posed."

On the way that explanations of physical facts can be distinctively mathematical see Lange (2013, p. 487) who suggests that "[d]istinctively mathematical explanations are 'non-causal' because they do not work by supplying information about a given event's causal history or, more broadly, about the world's network of causal relations. Such an explanation works instead by (roughly) showing how the fact to be explained was inevitable to a stronger degree than could result from the causal powers bestowed by the possession of various properties. If a fact has a distinctively mathematical explanation, then the modal strength of the connection between the causes and effects is insufficient to account for that fact's inevitability."

Batterman (2010) provides an ingenious account of the explanatory role of mathematics in empirical science without appealing to mathematical *entities* (for example numbers of graphs) playing an explanatory role, but focusing instead on mathematical

and equilibrium explanation (Sober 1983 Rice 2015, sec. 4). If all these are accepted as genuine cases of explanation, then it is definitely wrong to insist on including causal arguments in every explanatory account.[4]

operations and how they are involved in explanations of physical phenomena. What is of more immediate importance to the discussion in the text is the following position of Batterman (2010, p. 2): "But, it seems to me that there are very good reasons to deny that all physical explanations are causal explanations. The main reason is that if one pays attention to explanations offered by physicists and applied mathematicians, it is very often the case that one finds no appeal to causes at all. In fact, in many instances the various causal details need to be eliminated in order to gain genuine understanding of some phenomenon or other."

Bueno and French (2012) show how their inferential conception of the application of mathematics to physical phenomena is able to accommodate the "asymptotic reasoning" of Batterman. They show how partial mappings from mathematical structures to empirical phenomena can be established and ask (2012, p. 97): "[W]hat is doing the explanatory work? The inferential conception on its own does not settle this issue. It all depends on how the relevant mathematics that is used to represent the empirical set up is interpreted. That mathematics can be interpreted in realist terms or not, and it can also be interpreted as playing an explanatory role or not [...] Furthermore, those who hold that mathematics *does* play such an explanatory role owe us an account of the nature of explanation involved in the relevant examples of scientific practice Expressing it as neutrally as possible, any such account must be able to tell us how the mathematics and the relevant physical phenomena are related in a manner that goes beyond the representation of this relation via deduction or other formal devices. One option would be for such an account to say how it is that the relevant physical phenomena is brought about. One doesn't always have to appeal to causal factors in explicating this bringing about – one might draw on certain structural features, for example."

A central issue in the debate of the applicability of mathematics to empirical phenomena concerns the explanation of *patterns*. The first to provide a discussion of patterns as explananda was, as far as I can tell, Hayek in section 7 of his *Degrees of Explanation* (1955) and more explicitly in his *Theory of Complex Phenomena* (1967).

On the problematic of the application of mathematics in scientific practice see also Baker (2012) and Pincock (2012, 2015). Skow (2015) considers and argues in favour of the antipode, i.e., the possibility that physical arguments can explain mathematical "phenomena".

[4]   A very simple example of a non-causal explanation is provided by Braine (1972, p. 144): "[W]e produce a computation in order to explain why a mother cannot but fail in her attempts to distribute 23 strawberries equally and each undivided amongst her 3 children."

Other examples are provided by Lange (2013, p. 494): "Although an explanation that fails to identify causes may still be a causal explanation, not all explanations are causal. For example, that Samuel Clemens and Mark Twain are identical explains (non-causally) why they have the same height, birth dates, and so forth. For scientifically more prominent examples, consider some of the explanations in physics that appeal to symmetry principles. Energy conservation is explained by temporal symmetry: every law of nature follows from laws that are invariant under arbitrary time shift (i.e. under

Consequently, explanatory pluralism is different than the explanatory ecumenism that Jackson and Pettit (1992) endorse insofar as it is independent of any type of commitment to causality. Their position is a dual one since they adopt a fundamentalist attitude on matters of causality and an ecumenical one on matters of explanation. On the one hand, they argue that micro-level explanations and macro-level explanations very often, if not always, provide different kinds of information, information that would be lost if we made the methodological decision to always prefer micro-level, small-grain explanations over macro-level, large-grain accounts. The same would be the case if we decided to give up longer-grain explanations in favour of closer-grain stories. As it may be often beneficial to hold onto accounts of things that invoke high-level macro-factors, so it will be of explanatory benefit to stick with accounts that refer to temporally remote conditions. Explanations at different levels as explanations at different removes in time may provide different sorts of equally valuable information leading to the phenomenon to be explained (p. 18f.).

Though this explanatory ecumenism seems to be prima facie liberal, it goes hand in hand with another position which fundamentally restricts its resonance. They endorse a causal-information theory of explanation and give it a fundamentalist turn, arguing for a supervenience connection between the micro and macro level: that the macro is the way it is in virtue of how things are at the macro.[5] "Causal fundamentalism", that is, the position that every explanation must provide information on the causal history of what is to be explained, need not be an ingredient of explanatory pluralism.[6]

---

the transformation t→ t ± a for arbitrarily temporal interval a)." See also the useful discussion of non-causal explanations provided by Reutlinger (2014). Saatsi and Pexton (2013) show that Woodward's account can yield non-causal explanations by virtue of exhibiting only how the explanandum counterfactually depends on the explanans.

[5]   For good critical discussions of the approach of Jackson and Pettit see Weber and van Bouwel (2002, esp. sec. 6.3) and Mäki (2002, esp. sec. 3–5).

[6]   Jackson and Pettit call "causal fundamentalism" a stronger position, i.e., the "doctrine that properties, in particular causal-relational properties, at higher levels are supervenient on properties at lower levels in such a way that what happens at micro-levels determines what happens at macro" (p. 19). However, even the weaker position, i.e.,

Quite to the contrary, Lewis's dictum that "to explain an event is to provide some information about its causal history" (1986, p. 217), which Jackson and Pettit endorse, need not be followed: a genuinely pluralistic position can and should make as few commitments as possible to causality (and other metaphysical issues). Causal fundamentalism is an unnecessary restriction imposed on the process of producing successful explanations in science.[7]

4.2 EXPLANATORY PLURALISM IS INDEPENDENT OF THE VALIDITY OF REDUCTIONISM

Explanatory pluralism need not hinge on positions and discussions on reductionism. In the few cases that such a position is defended in the literature, be it in the philosophy of physics,[8] the philosophy of psychology,[9] the philosophy of biology[10] or the philosophy of the

---

that every explanation must provide information on the causal history of what is to be explained, is problematic as it is argued in the text.

[7] See the interesting remarks on the issue in Kitcher (2001, pp. 74–76). Skow (2014) also endorses a version of causal fundamentalism, excluding, however, 'in-virtue-of' explanations (p. 446f.): "I want to exclude one particular kind of explanation of events: 'in-virtue-of' explanations. Suppose someone asks: 'Why is the distance between A and B five meters?' One candidate answer is: because the shortest path from A to B is five meters long. This is (or at least is trying to be) an explanation. It looks like an attempt to explain why some fact obtains by citing some other fact or facts that 'ground' the target fact, that are the 'deeper' facts 'in virtue of which' the target fact obtains. In-virtue-of explanations are obviously non-causal. So the thesis that all explanations of events other than in-virtue-of explanations are causal remains an interesting and controversial thesis; it is this thesis that I will defend."

[8] See, e.g., Weatherall (2011).

[9] See, e.g., McCauley and Bechtel (2001), McCauley (2009, 2013) and Abney et al. (2014).

[10] See, e.g., Sterelny (1996). Grantham (1999) also advocates an explanatory pluralism in paleobiology, but his argument does not hinge entirely on a reductionist strategy. Plutynski (2004, p. 1211) endorses explanatory pluralism in population genetics focusing on a different aspect: "We are mistaken when we look to classical population genetics exclusively to explain particular events or states of affairs in the world. It's better to view population genetics as either providing explanations of classes of events or processes, or providing proofs of possibility. Thus, there are several virtues in which a theory or model may serve as explanatory; it may provide us with a novel Weltanschauung or 'open up the black box', or it may provide us with demonstration of possibilities. Such proofs may be of great relevance at certain stages of science." Fagan (2015) proposes "collaborative explanation" as a distinct way to account for mechanisms in biology, endorsing at the same time explanatory pluralism.

social sciences,[11] it has mostly been done in the context of the dispute between reductionists and anti-reductionists.[12] However, the passion for viewing scientific phenomena through the lenses of reductionism is a relic of logical positivism and of the epoch of the explanatory monarchy of physics (which designated all other sciences as "special") rather than of the explanatory liberal democracy of our times. Though the debate about the merits of the different varieties of reductionism is still continuing,[13] it is important to stress that one can be an explanatory pluralist without adopting a specific stance towards the issue of reduction. The description and normative appraisal of the provision of explanations is clearly an exercise in epistemology, so that ontological issues[14] can and should as far as possible not be the focus of the enterprise.[15]

[11]   See, e.g., Marchionni (2008).

[12]   Patrick Suppes, who was one among the first to issue a manifesto for pluralism in his presidential address to the Philosophy of Science Association in 1978, has discussed unity and reductionism concurrently, shaping, thus, the subsequent discussion (though his contribution was differentiated enough to include the other two long-standing themes of certainty of knowledge and completeness of science in making his case for the plurality of science). See Suppes (1978, pp. 9ff.). For a later manifesto for plurality of methods see Stump (1992). For a discussion of different versions of reductionism and how they are supposed to relate to explanatory pluralism see Steel (2004). For a discussion of levels of explanation see Potochnik (2010) and Gervais (2014) and of explanatory depth see Weslake (2010). For an excellent concise discussion of explanatory pluralism which I share see Godfrey-Smith (2003, pp. 197ff.). See also Thalos (2002) and Díez et al. (2013).

[13]   See for example Papineau (2009) and the critique of Schulman and Shapiro (2009).

[14]   Starting with ontological claims, which supposedly lend authority to particular strategies of explanation, has now become widespread. In philosophy of economics it has been recently rediscovered by Tony Lawson (1997, 2003) who has argued (among others) for a need of a social ontology which would supposedly solve all epistemological and methodological problems of the discipline of economics now still dominated by the neoclassical research programme. The "proof" of the existence of specific kinds of structures in the social realm is provided by transcendental arguments which are controversial. I cannot go into details here, but see van Bouwel (2004, p. 304) who speaks in this context of an "ontological fallacy" and remarks: "Lawson's way of reasoning about the relation between social ontology, on the one hand, and epistemology and methodology, on the other hand, is a recurrent phenomenon in the philosophy of the social sciences. It starts with certain a priori or necessary truths concerning social ontology, be justified by 'metaphysical commonplaces', (questionable) transcendental arguments or political convictions and, then, the methodological consequences seem to follow 'automatically' from the ontological stance." See also the discussions in Fullbrook (2008).

[15]   Besides, scientific understanding will not be the focus of the enterprise either. The recent revival of the notion of understanding in the literature of scientific

More generally, one can and should try to remain agnostic and, thus, neutral as far as this is possible with respect to metaphysical issues. This translates into the adoption of minimalist metaphysical commitments (which is not identical with the claim that metaphysical questions are not relevant at all[16]). As will be shown in some detail later in the book, my position of explanatory pluralism is thoroughly independent on what kind of metaphysical claims one might want to make with respect to the ontological structure of reality.[17]

More specifically, the requirement of viewing scientific phenomena and scientific laws in a *vertical way* based on the ontological thesis that all that exists is physical is not stronger than the requirement of viewing scientific phenomena and scientific laws in

---

explanation (e.g., Wilkenfeld 2014, Smith 2014) seems to ignore and/or neglect the age-old controversy on *Verstehen vs. Erklären* (see Mantzavinos 2005, ch. 4) and as Khalifa (2012 with further references) points out, introducing the notion of scientific understanding in the debate does not seem to have been very fruitful up to now.

[16] This is what Morrison seems to imply in summarizing her argument in her *Unifying Scientific Theories* in unsurpassable clarity (2000, p. 237):

"My general conclusion is that it is a mistake to structure the unity/disunity debate in metaphysical terms; that is, one should not view success in constructing unified theories as evidence for a global metaphysics of unity of disunity. Whether the physical world is unified or disunified is an empirical question, the answer to which is both yes and no, depending on the kind of evidence we have and the type of phenomena we are dealing with. In some contexts there is evidence for unity, whereas in others there is not. But nothing about that evidence warrants any broad, sweeping statements about nature 'in general' or how its ultimate constituents are assembled. The construction of theories that unify phenomena is a practical problem for which we only sometimes have the appropriate resources – the phenomena themselves may simply not be amenable to a unified treatment. Seeing unity in metaphysical terms shifts the focus away from practical issues relevant to localized contexts toward a more global problem – erecting a unified physics or science – one that in principle we may not be able to solve because nature may not wish to cooperate. The methodological point I want to stress is that eliminating metaphysics from the unity/disunity debate need not entail a corresponding dismissal of philosophical analysis. [...] Uncovering the *character* of unity and disunity is a philosophical task, one that will contribute to the broader goal of better understanding the practice of science itself. And given the progress and methods of empirical science, there is perhaps nowhere that metaphysics is less helpful." See also Woody (2015, p. 84) and the commentary by Love (2015, p. 91).

[17] My position of explanatory pluralism is, thus, thoroughly independent on such positions as Dupré's (1993) promiscuous realism, Cartwright's Dappled World (1999, 2007) or Hacking's Historical Ontology (2002), which defend metaphysical positions that are one way or another pluralistic.

a *horizontal way* – not on any a priori grounds. Adopting a "top-down" approach from institutions, relationships, feelings and other such entities to physical entities, forces and processes might be as productive as adopting a "side-to-side" approach when the aim is to provide successful explanations. One should remain agnostic about reduction: explanatory pluralism neither denies nor presupposes the metaphysical view that all macro-level phenomena (i.e., entities, events and processes) are just complex combinations of micro-level phenomena.[18]

Consider the problem of activities undertaken in heaven: Michael the archangel, when he was disputing with the devil about the body of Moses, did not dare to bring a slanderous accusation against him, but said, "The Lord rebuke you!" (Jude verse 9). We know from the Bible that the angels engage in different kinds of activities: they are instruments of God's judgements (Revelation 7:1; 8:2); they bring answers to prayer (Acts 12:5–10); they aid in winning people to Christ (Acts 8:26; 10:3); they observe Christian order, work and suffering (1 Corinthians 4:9; 11:10; Ephesians 3:10; 1 Peter 1:12); they encourage in times of danger (Acts 27:23–24); they care for the righteous at the time of death (Luke 16:22). Now, the critical question is the following: Is the ontological structure in heaven a vertical, hierarchical one, since the term "archangel" refers to Michael, the leader of the angels, or is it rather a horizontal one, Michael being as an archangel one among equals? This question can only be answered by a closer study of the source of all our knowledge with respect to the activity in heaven, that is, the Bible, and its answer is independent on whether angels are male or female.

Despite our high esteem for scientific knowledge, the situation remains analogously desperate. Instead of the Bible we have the corpus

[18]   Grantham (2004) argues strongly against the "Unity as Reduction" view which seems to prevail in contemporary discussions. He focuses instead on interconnection between fields rather than reduction between theories, which has the advantages of paying the due attention to the prevailing practices apart from the theories and of making unity a matter of degree. He remarks: "Even when reduction fails, the concepts and ontologies of the theories may be closely connected and the two theories can be strongly interdependent with regard to their heuristics, methods of confirmation, and explanations" (p. 140).

of scientific knowledge, but this might reveal to us as many details of the ontological structure of the world as the Bible does with respect to heaven. The epistemology of explanatory practices cannot hinge upon the prior answering of the question whether this and similar analogies are valid or not – since this is something that we might never be able to discover.

4.3 A NONAPOLOGETIC STANCE TO EXPLANATORY PLURALISM

Stating explanatory pluralism in a positive way is, of course, much harder and must avoid two dangers: first, of serving a merely *apologetic function* by providing ex-post legitimization of any and/or all explanatory scientific practices. Provided that there are indeed diverse kinds of explanatory practices in any domain of science, an explanatory plural*ity* in the sciences is the diagnosis of the prevailing state of affairs. In other words, plurality is a feature of the state of inquiry in a number, if not indeed all, areas of scientific research. The ways that plurality is characteristic for a particular area of inquiry can vary due to the different sources of plurality including the complexity of the phenomena (whether associated with crossing levels of organization or multiple factors within the same level of organization), the variety of explanatory interests[19] and the limitations of particular explanatory strategies vis-à-vis the phenomena (Kellert *et al.* 2006, p. xvi).

---

[19] In a recent discussion on how to account for the need of explanation – that is, *either along psychological lines* as Grimm (2008, p. 493f.) suggests that "a situation stands in need of an explanation for someone in virtue of the person's sense that there are various alternative ways the subject of the situation (a system, say, or a substance that constitutes the 'A' in a fact such as A is F) might have been. It is thus not the obtaining of a given fact per se, but instead the larger situation – the larger possibility space consisting of both fact and foil – that stands in need, or fails to stand in need, of explanation for someone", *or along normative lines* as Wong and Yudell (2015, p. 2883) suggest: "According to the map account, there is a real distinction between phenomena that need explanation and those that do not: a phenomenon needs explanation when it does not fit the map we are using, and whether it fits the map is a matter of fact independent of our assessment" – the implicit consensus is that there is a plurality and variety of explanatory interests.

The issue is what kind of stance one wants to adopt *about* this state of affairs. Plura*lism* is a view about this which has two facets.[20] The descriptive facet provides the very description of this state of affairs and seems relatively unproblematic:[21] it is very often the case that at any moment of time a plurality of explanatory practices prevail in any domain of science.[22] The normative facet is the crux of the matter: engaging in the process of discovering reasons in order to legitimize this plurality might degenerate into an apologetic exercise. This danger must be avoided and the key is to develop a position of explanatory pluralism that does not only save the phenomena of the plurality and variety of explanatory practices, but allows instead the use of critical resources.

A second, similar but distinct danger consists in sliding into an *anarchic position* which christens as acceptable any explanation offered in scientific discourse by entirely giving up the authority of any kind of scientific rationality.[23] In my view, however, philosophy

[20] I am only concerned here with explanatory pluralism and I therefore tackle the issue strictly in terms of explanatory practices. For a wider, extremely valuable approach called *integrative pluralism*, which also includes explanatory pluralism, see Sandra Mitchell (2003, part II, 2009a, ch. 6, and 2009b, and the important comment by Alt (2009)). Longino's (2013) ineliminable pluralism is also a general philosophical approach to pluralism of a wider scope. For an economic approach to philosophy of science embracing pluralism in science see Lütge (2001, 2004). See also Ruphy's "foliated" pluralism (2011).

[21] The stress is on *relatively* unproblematic. Morrison (2000, pp. 1ff.) has challenged the (what has now become the standard) view that when we look at scientific practice we see overwhelming evidence for disunity and opposes to the altogether banishment of unity. For a critical discussion of the many different ways that the general claim in favour of the unity of science has been historically defended see Richardson (2006, pp. 3ff.).

[22] This has been also supported by evidence presented in a recent pioneering study of experimental philosophy which shows – using text mining methods and analysis of random samples from data from the leading journal *Science* – that explanation is important in science, it is general across the scientific disciplines and it is a goal of scientific practice (Overton 2013).

[23] This is, in a sense, a radicalization of the first danger: fundamental relativism as the strategy of capturing the prevailing plurality in scientific (and non-scientific) explanatory discourse. For the dangers of relativistic pluralism see the interesting discussion in van Bouwel and Weber (2008, sec. VIII). For a discussion of the distinct danger of scepticism, i.e., the position that there is no such thing as distinctively explanatory information, see Nickel (2010, pp. 307ff.).

of science has a normative function to serve, so that a philosophical theory of explanation which intends to be pluralistic must clearly accommodate normative considerations. Working out the normative dimension of explanatory pluralism is clearly the main desideratum under the proviso that any kind of normative considerations cannot be formulated *in abstracto*, but must be founded on the existing scientific practices.

I propose that these two dangers can be evaded if one theorizes in terms of explanatory games. Let us proceed with the elaboration of this claim.

5   The Explanatory Enterprise

Explanatory activity is undertaken by imperfect biological organisms with a limited cognitive capacity in interaction with artefacts in a specific social context. The explanatory enterprise is a social process, and it consists of the attempt of the participants to this process to provide answers to puzzles and solutions to theoretical problems. The explanatory enterprise is embedded in certain practices employed by the participants and unfolds according to normative standards that have emerged in a long evolutionary process of trial and error. There is a history to the explanatory enterprise, a history that includes the more and less successful attempts of answering specific "Why?"-questions, the development of more and less accurate means of representation devised to answer such questions and the permanent change of institutional constraints under which the participants engage in their activities. The explanatory enterprise has a prehistory rooted in the attempt of our remote ancestors to understand what was going on in their immediate physical and social environment and a history reflected in the progressive exposure of explanations to a critical attitude and testing. Mythical explanations of rain caused by a frustrated Zeus sitting on the top of Olympus throwing thunder to the mortal have given place to scientific explanations of natural phenomena tested in different experimental settings with the aid of highly sophisticated techniques. The explanatory enterprise emerges as a human phenomenon – therefore permanently unfinished – and it should be understood as a *project* in which humanity has been engaged for most of its history.[1]

---

[1]   For a similar argument in the realm of ethics, see Kitcher, *The Ethical Project* (2011, p. 2).

FIGURE I Elena's "Why-Question"

5.1   WARUM REKNETZ?

Elena, a six-year-old girl living in Germany, has observed the phenomenon of rain and asked herself why it rains. She has used the means of
representation that she has been taught in the first year of her elementary school, that is, letters combined in words and short sentences,
drawings and colours, in order to formulate the question "Why does it
rain?" and to answer it. The explanation is stated hypothetically since
it starts with "I believe." There is a meta-theoretical remark on the
third page: she writes that "this story ends here". Beyond the line it is
stated that there a new story begins.

The question, the answer and the meta-theoretical remark are
all full of errors.

Figures 1–3, each on one page, with the question, the answer and
the meta-theoretical remark, encapsulate in an archetypal way the
main ideas of the book, as it will become clear later.

5.2   THE COGNITIVE DIVISION OF EXPLANATORY LABOUR

At every moment of time there is a stock of explanations available in
a society proposed by ordinary people "in the wild" or by specialists

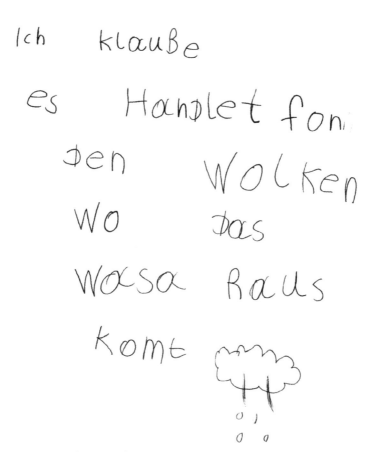

FIGURE 2 Elena's explanation

organized formally or semi-formally within specific organizational structures such as churches, universities, etc. This explanatory reservoir is distributed among diverse individuals and groups in the society under conditions of a cognitive division of labour. The terms of provision, control and dissemination of explanations in this collective explanatory enterprise are regulated by the different rules that the participants have come to adopt over time. These rules incorporate the normative standards that guide the processes of discovery and justification of explanations as well as the modes of their

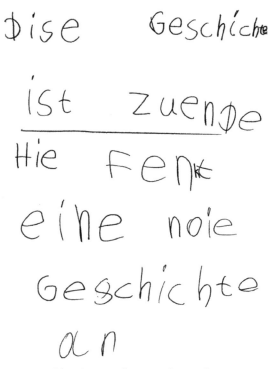

FIGURE 3 Elena's meta-theoretical remark

communication, dissemination and adoption. They constitute *the rules of the explanatory game* that the participants are playing.[2]

---

[2]   There is a long philosophical tradition in the use of the notion of a game. The use of the notion of "παιδιά" is prevalent in Attic Greek. It is characteristic that in Plato's *Laws* the whole inquiry into the subject of laws is described by the Athenian as "playing a sober game suitable to old men" (περί νόμων παίζοντας παιδιάν πρεσβυτικήν σώφρονα (685a)). To my knowledge, it is the use by Epictetus in his *Discourses* that has had a considerable influence in modern discussions. It is preserved by Arrian in Bk. II, ch. 5, where the Greek words "σφαιρίζειν" and "παίζειν" are included in the following context (in the original Greek and in the English translation):

τοιγαροῦν Σωκράτης ἤδει σφαιρίζειν. πῶς; παίζειν ἐν τῷ δικαστηρίῳ. "λέγε μοι," φησίν, "Ἄνυτε, πῶς με φῂς θεὸν οὐ νομίζειν; οἱ δαίμονές σοι τίνες εἶναι δοκοῦσιν; οὐχὶ ἤτοι θεῶν παῖδές εἰσιν ἢ ἐξ ἀνθρώπων καὶ θεῶν μεμιγμένοι τινές;" ὁμολογήσαντος δὲ "τίς οὖν σοι δοκεῖ δύνασθαι ἡμιόνους

Explanations emerge in the process of playing the game. How the game is played depends on the rules of the game. Obviously, different people play different explanatory games which give rise to different outcomes. The distinction between rules and actions within rules is

μὲν ἡγεῖσθαι εἶναι, ὄνους δὲ μή;" ὡς ἁρπαστίῳ παίζων. καὶ τί ἐκεῖ ἐν μέσῳ ἁρπάστιον τότ᾽ ἦν; τὸ δεδέσθαι, τὸ φυγαδευθῆναι, τὸ πιεῖν φάρμακον, τὸ γυναικὸς ἀφαιρεθῆναι, τὸ τέκνα ὀρφανὰ καταλιπεῖν. ταῦτα ἦν ἐν μέσῳ οἷς ἔπαιζεν, ἀλλ᾽ οὐδὲν ἧττον ἔπαιζεν καὶ ἐσφαίριζεν εὐρύθμως. οὕτως καὶ ἡμεῖς τὴν μὲν ἐπιμέλειαν σφαιριστικωτάτην, τὴν δ᾽ ἀδιαφορίαν ὡς ὑπὲρ ἁρπαστίου. δεῖ γὰρ πάντως περί τινα τῶν ἐκτὸς ὑλῶν φιλοτεχνεῖν, ἀλλ᾽ οὐκ ἐκείνην ἀποδεχόμενον, ἀλλ᾽ οἵα ἂν ᾖ ἐκείνη, τὴν περὶ αὐτὴν φιλοτεχνίαν ἐπιδεικνύοντα. οὕτως καὶ ὁ ὑφάντης οὐκ ἔρια ποιεῖ, ἀλλ᾽ οἷα ἂν παραλάβῃ περὶ αὐτὰ φιλοτεχνεῖ. ἄλλος σοι δίδωσι τροφὰς καὶ κτῆσιν καὶ αὐτὰ ταῦτα δύναται ἀφελέσθαι καὶ τὸ σωμάτιον αὐτό. σὺ λοιπὸν παραλαβὼν τὴν ὕλην ἐργάζου. εἶτα ἂν ἐξέλθῃς μηδὲν παθών, οἱ μὲν ἄλλοι ἀπαντῶντές σοι συγχαρήσονται ὅτι ἐσώθης, ὁ δ᾽ εἰδὼς βλέπειν τὰ τοιαῦτα, ἂν μὲν ἴδῃ ὅτι εὐσχημόνως ἀνεστράφης ἐν τούτῳ, ἐπαινέσει καὶ συνησθήσεται· ἂν δὲ δι᾽ ἀσχημοσύνην τινὰ διασεσωσμένον, τὰ ἐναντία. ὅπου γὰρ τὸ χαίρειν εὐλόγως, ἐκεῖ καὶ τὸ συγχαίρειν.

In that sense Socrates knew how to play the game. "What do you mean?" He knew how to play in the court. "Tell me, Anytus", said he, "in what way you say that I disbelieve in God. What do you think that divinities are? Are they not either children of the gods, or the mixed offspring of men and gods?" And when Anytus agreed, he said, "Who then do you think can believe in the existence of mules and not in asses?" [Plato, *Apology*, 27c] He was like one playing at ball. What then was the ball that he played with? Life, imprisonment, exile, taking poison, being deprived of his wife, leaving his children orphans. These were the things he played with, but none the less he played and tossed the ball with balance. So we ought to play the game, so to speak, with all possible care and skill, but treat the ball itself as indifferent. A man must certainly cultivate skill in regard to some outward things: he need not accept a thing for its own sake, but he should show his skill in regard to it, whatever it be. In the same way the weaver does not make fleeces, but devotes himself to dealing with them in whatever form he receives them. Sustenance and property are given you by Another, who can take them away from you too, yes and your bit of a body as well. It is for you, then, to take what is given you and make the most of it. Then if you come off without harm, others who meet you will rejoice with you in your safety, but the man who has a good eye for conduct, if he sees that you behaved here with honour, will praise you and rejoice with you: but if he sees a man has saved his life by acting dishonourably, he will do the opposite. For where a man can rejoice with reason, his neighbour can rejoice with him also.

Adam Smith in *The Theory of Moral Sentiments* (1759/1976, Part VII, Sect. II, Ch. I, p. 278f.) writes: "Human life the Stoics appear to have considered as a game of great skill; in which, however, there was a mixture of chance, or of what is vulgarly understood to be chance. In such games the stake is commonly a trifle, and the whole pleasure of the game arises from playing well, from playing fairly, and playing skillfully." Adam Ferguson in his *Principles of Moral and Political Science* notes (1792, vol. I, p. 7): "The Stoics conceived human life under the image of a Game; at which the entertainment and merit of the players consisted in playing attentively and well, whether the stake was great or small."

The notion of the game has been employed more recently in philosophy by Popper in his *Logik der Forschung* (1934, Kap. 11): "Wir betrachten die methodologischen Regeln als Festsetzungen. Man könnte sie die Spielregeln des Spiels >empirische Wissenschaft<

constitutive of the analysis of the explanatory enterprise in terms of games. The players of a respective game, be they ordinary people, scientists or religious preachers, are constrained at any moment of time by a series of rules concerning their explanatory activities. The rules of the game divide, in principle, the innumerable possibilities of providing explanations into those that can be undertaken and those that cannot. The rules of the game define the means that the players can use to represent the phenomena that they want to explain as well as the class of explanatory strategies that the players are allowed to undertake. They structure the interaction between the players and shape the way that the explanatory game is played over time. In the games that scientists are playing, call them *scientific explanatory games*, one set of rules defines, for example, the requirements of logical consistency that explanations must respect. Such rules are not characteristic of other kinds of games; for example, they do not apply to the explanatory games that participants of diverse religious confessions are playing, call them *religious explanatory games*, where miracles and paradoxes are acceptable explanatory strategies. The respect for the rules of logical consistency gives rise to another type of explanatory process than when these kinds of rules are not respected; and it leads, of course, to different outcomes.

This view conceptualizes the explanatory enterprise as a set of explanatory games and highlights explanatory activity as fundamentally social. In the case of science, for example, there is a cognitive division of labour: some provide the big intuitions; others build the bridges to existing theoretical structures, providing

---

nennen. [...] ähnlich können wir die Untersuchung der Regeln des Wissenschaftsspiels, der Forschungsarbeit, auch *Logik der Forschung* nennen." Wittgenstein has made extensive use of the notion of "Sprachspiele" in his *Philosophische Untersuchungen* (1953). The notion of game is at the centre of formal game theory originating in *The Theory of Games and Economic Behavior* (1944) by von Neumann and Morgenstern. In political philosophy and in the social sciences the notion of game has been employed most prominently by the Nobel Laureates Friedrich A. von Hayek, notably in the second volume of his *Law, Legislation, and Liberty* (1982), James M. Buchanan, most clearly in his (co-authored with Geoffrey Brennan) *The Reason of Rules: Constitutional Political Economy* (1985), and Douglass C. North, more influentially in his *Institutions, Institutional Change and Economic Performance* (1990).

the unification of phenomena; others elaborate on the means of representation; and still others work out the empirical testing and applications to different range of phenomena. Explanations are the outcome of the complex processes of social interaction between individuals pursuing different aims and undertaking different tasks. Even in those cases where explanations deemed to be very successful seem to be prima facie due to the labours of single scientists, it is never the case that she masters all the practices *and* has invented all means of representation *and* has designed all techniques necessary for the provision of them. In other words, it is hardly ever the case that all parameters of the explanatory process are controlled by one individual and all the ingredients of a successful explanation can be offered by a single scientist.

5.3   THE EXPLANATORY PROCESS

A conceptualization of activities in terms of explanatory games goes hand in hand with a focus on explanation as a process rather than as an outcome. The task is to highlight the *complex process of explanation* rather than to pose the issue as if explanation were a *static situation*. During an evolutionary process of trial and error explanatory activities are undertaken according to the prevailing rules of the game, and a permanent flow of explanations is produced, tested and retained or discarded. Novelty is a permanent feature of this process since new ways of representing the phenomena constantly emerge, new ways of testing are being constantly designed and new means of criticism are constantly invented. The traditional view of explanation in its many facets has failed to see this point and has undertaken scholastic efforts instead to analyse the outcomes of this process, highlighting mainly the logical aspects of what was supposed to be "the explanatory relation". There has been silence about a whole range of issues pertaining to the processes of the construction of (successful and less successful) explanations and to the diverse factors that play a role in their rise and demise – quite a natural attitude since the philosophical project has been concerned with the explication of the concept of explanation rather than with the analysis of the process of explanatory activity. More fatally still, the hidden consensus has been that there is a single outcome, "the explanation" to be described, rather than a range of explanations produced in a constant process; and it has been thought that there is a single ideal, "the successful explanation" that is to be normatively appraised, rather than a great range of normative rules that structure the process, which are to be criticized and refined.

To use a metaphor, the traditional view of explanation has focused on what comes out of the pipe instead of the flow in it; and it has offered the analysis of the drain that comes out of the flow as if it were analogous to the pressure of the influx at the beginning of the pipe. Between the two ends, however, lies an elastic and changeable reservoir, whose magnitude depends on a series of circumstances. Analogously, in the case of ordinary explanations the analysis should focus on the flow of the river in the valley rather than on how the landscape around the mouth of the river is shaped by the water ending in it. One can grasp the explanatory process if one conceptualizes it as a continuous flow or river, which produces unceasingly in its mouth the final products emerging in multiple transformative processes after the initial insertion of inputs. At every moment of time many such streams flow or, even more, complexly branched river systems emerge next to each other, each a bit more advanced than the next. The final products of all those streams appear in more or less distant points in time.

The flow of explanations is channelled by the rules that govern the explanatory activity. Depending on the rules, only certain classes of activities can be undertaken; consequently a whole range of explanations that are possible in principle are never produced and never considered. The rules of the game are whatever constrains the flow and gives it a shape; there can be natural or artificial impediments that channel the flow in different ways: little stones and branches which push the flow in a certain direction or big artificial dams which interrupt it violently and force the water to collect in artificial lakes. The rules of the game influence decisively the explanatory process and shape the possible final outcomes.

Explanatory activities take place in historical time. The history develops according to basic rules that set the trajectory for development and which the participants learn in their socialization process. The historicity of explanations is embedded in the rules of the explanatory game: in the case of scientific explanations, this occurs in processes whereby *apprentices* learn the ways to play the game from *veterans* in the respective scientific community (Kitcher, 1993, pp. 58ff.); in the case of religious explanations, it occurs in the process whereby *deacons* learn the rules of the explanatory game from *priests* in the respective religious community. The historicity of the explanatory enterprise refers to nothing more than the learning history of the individual members of the respective group regarding which explanatory strategies are permissible and which are not. And the best way to analyse the slogan "history matters" is to analyse the rules that govern the respective explanatory games. How should such an analysis proceed and what can it accomplish?

6 The Rules of the Explanatory Game

The analysis of the rules of an explanatory game can be of both a descriptive and a normative nature. I will tackle the descriptive task in this and the next three chapters before taking up the normative task in the subsequent two chapters.

A precise description of the rules of a game not only includes the way that the rules have emerged over time but also their precise content. It is straightforward that there are different types of explanatory games, and the means for classifying them refer to the similarity of the rules that govern the explanatory activities of the respective participants. There are religious explanatory games, mythical explanatory games, scientific explanatory games, etc. The description of the different kinds of games requires the exposition of the special rules that characterize them and an account of how they differ between them. In the game that the members of the Christian community are playing, for example, a great range of phenomena is explained by following rules that presuppose a religious experience of revelation. The doctrine of the Trinity is not an explication of the notion of God or, more neutrally, a definition of God, since the Christian approach is neither rationalistic nor analytical; rather, it fundamentally derives from an intimate encounter during which God is perceived to be *revealed* as Father, Son and Holy Spirit. The three hypostases are united in one Godhead, in Being, though each hypostasis is distinct. In every action in which God engages, all three hypostases work together as one. The application of the doctrine of the Holy Trinity gives rise to explanations of phenomena such as rain, wars, etc. that are impossible in explanatory games that are governed by other sets of rules.

In what follows, I will not undertake the task of describing different kinds of games by way of developing a typology or classification scheme, but will devote my attention instead to the harder task of providing an abstract characterization of an explanatory game. I will subsequently provide two case studies in order to exemplify it in more concrete terms.

6.1   AN ABSTRACT CHARACTERIZATION OF THE RULES OF THE GAME

An abstract characterization of the rules of the explanatory game can be given by distinguishing between four kinds of rules:

1. **Constitutive rules**. There is a basic set of rules that constitute an explanatory game as a game. They include three categories of rules: (a) *Rules determining what counts as an explanandum.* Aristotle famously regarded motion as an explanandum and he provided causes as an explanans. Newton did not regard motion as an explanandum. Motion was taken to be what it was, the explanandum being the change of motion. Aristotle and Newton played a different kind of explanatory game. (b) *Rules determining what must be taken as given.* Every game takes place within a context of background knowledge that remains unquestioned and is taken as given. There is a bedrock of unquestioned facts, beliefs and practices which are left out of the game, implicitly or explicitly. When August Cournot explained the pricing behaviour of a monopolist he took as given that the earth is round, all human beings can breathe and that the cooking of a pie requires a system that produces heat. (c) *Rules determining the metaphysical presuppositions.* No explanatory game can take place in a metaphysical vacuum. The metaphysical assumptions act as constraints on the generation of the other rules, and belong therefore to the constitutive rules. The structure of the game is predicated on prior assumptions concerning the way the world is and by what means it is explainable in principle. These rules can be implicit or explicit and they can vary from stone-age metaphysics to highly refined metaphysical assumptions.

2. **Rules of representation.** Because of the hospitality the people of Attica showed Demeter when she was searching for Persephone, the Olympian goddess first gave agriculture to the human race. This mythical explanation of the emergence of agriculture is not only represented by means of an oral story (later also documented in written form) using metaphorical language but also with the aid of some visual representations such as a fifth-century BCE red-figure attic stamnos, showing Triptolemos teaching agriculture, the gift of Demeter (Clark 2012, pp. 7ff.). In contemporary scientific explanatory games, representation occurs by means of artefacts, both concrete, such as graphs, scale models, computer monitor displays, etc., and abstract, such as mathematical expressions (van Fraassen 2008). Which rules of representation guide the explanatory activities is fundamentally important for the quality of explanations generated during the game.

The most common form of representation is linguistic representation. Texts consist of written words which are composed by letters, identifiable by their shapes, and both width and height are relevant. The meaning of linguistic statements is determined by the sequence of letters, spaces and punctuation and what is important is that the *spatial* form of serial representations is largely arbitrary with respect to their meaning; the shapes of letters are unrelated to the referents of the words and sentences they comprise. The fact that the visible forms of written sentences are arbitrary with respect to their meaning is the main source of their convenience and flexibility of textual representation. Linguistic representations provide an extremely easy format to express abstract or general thoughts (Perini 2005a, p. 264) and they act as the basis on which very diverse sets of inferential rules – including logical rules – can be put to use. Linguistic representations that are syntactically disjoint, that is, where each mark is assigned to at most one character and articulate, that is, where each can be finitely differentiated (Goodman 1976), include besides natural language, also symbolic logic and mathematical expressions.

Linguistic representations are serial representations whose form is arbitrary with respect to their referents and must be distinguished from visual representations which are symbolic systems whose conventional rules involve determining referents based on relations between the form of the symbol and properties of the referent (Perini 2005a, p. 269). Visual representations exhibit most importantly some spatial features that are interpreted to mean something about the phenomenon that they represent. Spatial relations of a visual representation can refer to *spatial* relations as in a diagram of a molecule, *temporal* relations as in the case of time lines or to relations between *properties* as in graphs (Perini 2005b, p. 914). Other visible features such as colour can also be used to represent aspects of a phenomenon.

However, there are important differences among visual representations. *Diagrams*, for example, have syntactic features that other visual representations, those that look more like pictures, do not. They are characterized by an articulate syntax so that a complete dissection of each diagrammatic representation to its component parts is possible. In addition to this type of syntax, the semantics of diagram systems assign referents to, in principle, unambiguously identifiable characters, so that the meaning of a diagram is a function of the meaning of the atomic characters along with the way that these are arranged. A diagram can thus be given a linguistic translation provided that the language contains terms for what the atomic characters refer to as well as terms for the relations that the spatial relations among the atomic characters refer to (Perini 2005c, p. 260). In principle then, the set of rules of a diagrammatic representation of a phenomenon can be translated to a set of rules of a linguistic representation of the same phenomenon. However, a serially formatted representation would not be concise (Teller 2008, p. 435), since in order to describe all the different kinds of relations a series of statements would be required and a long conjunction thereof. More importantly the visible form of such a linguistic representation bears no relation to the structure of the phenomenon it is supposed to depict. And although some linguistic and diagrammatic representations are informationally equivalent, this does not imply computational equivalence nor does it invalidate the fact that diagrams can be more efficient platforms for drawing inferences than informationally equivalent linguistic representations (Larkin and Simon 1987).

Pictorial representations, although also visual representations, are distinct from diagrams in that they avail of pictorial syntax and semantics and are not translatable, not even in principle, into linguistic representations. Their distinctive feature is that they do not have articulate syntax, since one cannot tell exactly what character a mark instantiates and they are dense systems, that is, there is a character ordered between any two characters (Perini 2005b, p. 916). Examples of such pictorial representations which cannot be translated into linguistic expressions include graphs, micrographs, etc. The great variety of pictorial representations is largely due to a variety in form–content relations. Particular images are only comprehensible if appropriate conventional interpretative rules of what one sees are followed, in order to relate the visible feature of a picture to what it represents. One must interpret grey colours differently in a black-and-white photograph, for example, than in a colour photograph. And of course pictures do not represent all aspects of their subject matter, but are selective about the properties they represent. Colour photographs, for example, represent the features of a specific object only from a specific angle, omitting both information about the visible features on the opposite side and about the object's non-visible features (Perini 2013, p. 274). And maps, to give a further example, are associated with two sorts of reading conventions which all competent map users grasp: one sort of conventions link the syntactic elements of the map to entities in nature and the other sort prescribes the way to connect the properties of the visual (sometimes coloured) display with properties of these entities; a map has, thus, a certain syntax and the set of reading conventional rules fixes its semantics (Kitcher and Varzi 2000, p. 377f.).

This discussion should have hopefully made clear that there is a great variety of possibilities to form representations (Frigg and Hunter 2010), the linguistic representations and visual representations being the more common and important ones (but there are, of course, also acoustic representations (Palmieri 2012), representations of taste and of odours). Representation in an essentially triadic relationship involving three relata: the representational-bearer, the representational object and the intepretant (or interpretation) in a cognitive agent (Files 1996, pp. 401ff., Giere 2006, p. 60). Representation is constituted by these three relata and the relations obtaining among them. Rules of representation determine how these relations are structured and include three categories of rules: (a) *Rules determining which entities count as representation-bearers.* Certain artefacts such as physical models, maps, pictures, mathematical expressions, etc. are created or designed for the express purpose of functioning as representation-bearers. Other artefacts which are not designed to serve as representation-bearers can be given the function of representing (Files 1996, p. 406). In both cases there are specific rules which single out these entities as representation-bearers. (b) *Rules determining by virtue of what a representation-bearer is supposed to represent.* Since the representation-bearers possess the property of "aboutness", as the most fundamental feature of representation, rules must determine by virtue of what a representation-bearer possesses the property of "aboutness" (Files 1996, p. 400, van Fraassen 2008, p. 27). Metal plates representing four bases along with rods arranged helically around a retort stand represent as a whole DNA (deoxyribose nucleic acid) in virtue of sharing the same shape and structure. (c) *Rules determining by virtue of what a representation-bearer is connected with the representational object,* that is, the object or state of affairs that a representation-bearer is about. These rules fill in all the details of how the connection is supposed to take place, for example, by denoting in the case of linguistic expressions, by resemblance in case of visual representations,[1] a combination of both or by entirely other ways.

3. **Rules of inference.** In every explanatory game a set of inferential strategies are used which aim at whatever is regarded as the explanandum. Inferences act on the representations provided and they can be of very different sorts (Suárez 2004, p. 777, Suárez 2015, pp. 45ff.). In religious explanatory games, for example, inferences are frequently made from experiences of revelation to phenomena easily

---

[1]   See Daston and Galison (2010, p. 382): "Representation is always an exercise in portraiture, albeit not necessarily one in mimesis. The prefix *re-* is essential: images that strive for representation present again what already is. Representative images may purify, perfect, and smooth to get at being, at 'what is'. But they may not create out of whole cloth, crossing over from nature into art."

Greenberg (2013) questions the validity of resemblance theory in visual representations. The fundamental intuition of resemblance theorists concerns the rejection of the supposed analogy between linguistic and pictorial representation since successful pictorial representation does not seem to be arbitrary at all: "The relationship between a drawing, photograph or perceptual representation of a scene and the scene itself is one of intimate correspondence, nothing like the stipulative association between a word and its denotation " (p. 216). Greenberg questions the analysis of resemblance theorists according to which the cornerstone of representation is accurate depiction which is "grounded in resemblance", developing an argument which is more familiar to artists and geometers than philosophers: accurate images are produced by following particular *transformation rules* for projecting three-dimensional scenes onto two-dimensional surfaces. Acknowledging the existence of many kinds of projection, corresponding to myriad systems of depiction, he shows that linear perspective can be characterized in terms of similarity, but that curvilinear perspective – which according to recent empirical research (Rogers and Rogers 2009) characterizes also human visual perception – is an intractable counterexample of resemblance theory. In other words, the cornerstone of contemporary resemblance theory which is *similarity* defined as *sharing properties* (isomorphism being a species of abstract similarity) can be retained in accounting for linear perspective projection as just one of infinitely many 3D-to-2D mappings, but not in accounting for *curvilinear perspective projection*. The latter projection *requires* "for accuracy that a picture *differs* from its subject according to specific rules of geometric transformation. Thus, [his] complaint with resemblance theory is not that there are no systems of depiction that can be grounded in resemblance, but rather, that there are some systems of depiction that cannot. [His] ultimate diagnosis is that resemblance theory has mischaracterized the basic architecture of accurate depiction. Rather than resemblance, accurate depiction in general is grounded in the more inclusive phenomena of geometrical transformation" (p. 263). Ultimately, "[t]he key feature of curvilinear perspective that undermines the analysis is that there is no way of preserving the shape properties of a flat surface under curvilinear projection" (p. 271). "Such considerations suggest that, in general, accurate depiction cannot be characterized wholly in terms of difference, or wholly in terms of similarity, but should instead be defined by the broader notion of *transformation* that incorporates both" (p. 282f.).

observable without sophisticated means of scientific representation. In modern scientific explanatory games, much more subtle and complex sets of rules of inference are normally used. Such rules can be very specific and very diverse, ranging from recipes to formulate predictions to instructions about how to construct arguments.

It is important to stress that the different types of representation allow of different kinds of inference. Linguistic representations which are characterized by an immense flexibility allow of a great array of inferences, ranging from very specific to very general inferential rules. In scientific explanatory games the respect of some formal requirements is of great importance. A vast array of transformations based on formal rules of inference are performed, especially on a specific subcategory of linguistic representations, that is, mathematical expressions.

It is a common error to implicitly or explicitly assume that arguments can be built only when specific rules of inference are carried out on linguistic representations. Since explanations are often arguments, this assumption would exclude the possibility of constructing arguments based on visual representations. Euclidean geometry is a nice counterexample which shows that rigorous inferences can be drawn on strictly visual representations. Euclid's first postulates are literally used as drawing instructions, such as "draw a straight line from any point to any point". The drawing rules specify the kinds of modifications one can make to a diagram, leading one from an initial diagram to a different final visual representation (Perini 2005a, p. 265f.). Such a use of inferences acting upon visual representations neatly shows that the construction of arguments does not depend on the availability of linguistic representations. What is moreover important is that there are specific types of rules of inference which can be carried out exclusively on visual representation and are largely enabled by such representations (Sheredos *et al.* 2013). Tables that arrange linguistic data in two dimensions, for example, *facilitate* inferences about higher-order relations between features of a set of values for one attribute and features of another set, *in virtue of the fact* that tables are virtual representations. Translation of the *content* of a table into a serial linguistic representation can account for the content of the table, but it fails to capture how the content of the table *relates to the conclusion the table supports* (Perini 2005b, p. 923). And there is a great diversity of rules of inference that act on pictorial representations and other representations specifiable in each case.

4. **Rules of scope.** These are rules of specification, that is, they give instructions about the scope of phenomena to which the explanatory game should be applied. The scope of the explanatory game that evolutionary theorists were engaged in originally included plants and animals. The scope has been then extended to include human culture. It has been extended and specified further to include even routines employed by firms in market competition and much more. The rules of scope are responsible for one main property of games, which I would like to call *nestedness*. They give instructions on how to dock one explanatory game into another and so provide nested games. The easier the fit between the rules of scope of different explanatory games, the higher the degree of their nestedness and the greater the potential of interlocking different explanatory games. For example, the rules of scope of the explanatory game of biological evolution and the rules of scope of the explanatory game of growth economics have turned out to be congruent, so that the economics of evolutionary change by Nelson and Winter (1982) has emerged as an explanatory game, nested in the two other games.

So, it is easier to get into a game if nestability is given. You need not learn (a lot of) new rules in order to get into it. (Nestability is, thus, in a sense the analogue of depth in the reductionist accounts of explanation.)

In a nutshell the rules of scope comprise the instructions about where to apply the explanatory practices of the game and how to apply the game to new phenomena – they are, thus, of a dynamic nature by virtue of extending or contracting the range of phenomena on which the specific explanatory practices of the game are to be applied.

Explanatory games are framed by *constitutive rules* and the explanatory activities unfold in a process during which *rules of representation, rules of inference* and *rules of scope* are combined and appropriately synthesized to generate explanations. But let us go to the scene of action, since, to recall Seneca, the way is long if one follows precepts, but short and efficacious if one follows examples. I will now turn to two case studies, one from economics and one from medicine.

6.2   A CASE STUDY FROM ECONOMICS

In order to exemplify further this rather abstract characterization of a game, I am going to discuss very shortly an explanatory game in the history of economic analysis that concerns the value of commodities. I will start with the explanatory game that classical political economists of the eighteenth century were playing and will then show how this game was transformed into the one that marginalist economists came to play at the end of the nineteenth century.

1. The *constitutive rules* of the explanatory game consisted first (a) in what counted as an explanandum. Adam Smith observed that

> [t]he word VALUE [...] has two meanings, and sometimes expresses the utility of some particular object, and sometimes the power of purchasing other goods which the possession of that object conveys. The one may be called "value in use"; the other "value in exchange". The things which have the greatest value in use have frequently little or no value in exchange; and on the contrary, those which have the greatest value in exchange have frequently little or no value in use. Nothing is more useful than water: but it will purchase scarce any thing; scarce any thing can be had in exchange for it. A diamond, on the contrary, has scarce any value in use; but a very great quantity of other goods may frequently be had in exchange for it.
>
> *(Smith 1776/1976, Bk. I, Ch. IV, p. 32f.)*

He then goes on to state that his aim is "to investigate the principles which regulate the exchangeable value of commodities" (Smith 1776/1976, p. 33). The value in exchange also remained the explanandum in the work of David Ricardo, as well as in that of John

Stuart Mill and other classical economists. Further (b) the explanatory game took place within a context of background knowledge, both natural scientific and common sense knowledge, which remained unquestioned. This bedrock of unquestioned facts and beliefs were left out of the explanatory game in an implicit manner. Finally, (c) the main rule regulating the metaphysical presuppositions was that the social world is in principle knowable, that it exemplifies an order commensurate to the order of nature and that it is governed by laws which are discoverable.

2. In order to explain the value of commodities the classical economists used natural language and numerical examples as the sole means of representation. The *rules of representation* that structured their explanatory game were, thus, very basic.

3. The *rules of inference* used to explain the exchange value of commodities were the rules of logic prevailing at that time, along with a series of law-like statements. David Ricardo famously stated in the preface of his Principles (1817/1951, p. 5) that "[t]o determine the laws which regulate this distribution, is the principal problem of Political Economy", and the explanation of value was supposed to be based on some of these laws. The most important one is included in the title of chapter I, section I of his classic work: "The value of a commodity, or the quantity of any other commodity for which it will exchange, depends on the relative quantity of labour which is necessary for its production, and not on the greater or less compensation which is paid for that labour" (Ricardo 1817/1951, p. 11).

4. The *rules of scope* for the game were formulated by Adam Smith. "In the early and rude state of society which precedes both the accumulation of stock and appropriation of land" the values are the exchange relationships between the goods reflecting the quantities of labour: "If among a nation of hunters, for example, it usually costs twice the labour to kill a beaver which it does to kill a deer, one beaver should naturally exchange for or be worth two deer" (Smith 1776/ 1976, Bk. I, Ch. VI, p. 53). In "improved societies", that is, "[a]s soon as stock has accumulated in the hands of particular persons" and "[as]

soon as the land of any country has all become private property" (Smith 1776/1976, Bk. I, Ch. VI, pp. 54ff.), the value of the commodities consists of the remunerations of labour (wage), capital (profit) and land (rent). Since this "early and rude state of society" is conjectural, the scope of the explanatory game is the "commercial society". An additional rule of scope is explicitly mentioned by Ricardo (1817/1951, p. 12): "In speaking then of commodities, of their exchangeable value, and of the laws which regulate their relative prices, we mean always such commodities only as can be increased in quantity by the exertion of human industry, and *on the production of which competition operates without restraint*" (my emphasis). In other words, the explanation of value is to be provided only in competitive markets, not on all markets – this is a serious limitation of the scope of this explanatory game. However, this very limitation enables this explanatory game to become embedded into other explanatory games, most prominently the explanatory game that has as an explanandum the value of the factors of production. Under competitive conditions the "whole annual produce of labour of every country, taken complexly, must resolve itself into the same three parts, and be parcelled out among different inhabitants of the country, either as the wages of their labour, the profits of their stock, or the rent of their land" (Smith 1776/1976, Bk. I, Ch. VI, p. 58f.). The explanatory game of the value of commodities is nested into the explanatory game of the value of factors of production, and this is made possible by the rules of scope, that is, the commercial society and the conditions of market competition. Under, say, monopolistic conditions, the nestedness is not given.

The explanatory game played by classical economists for approximately one century was structured by the sets of rules that I have indicated. Within these rules their explanatory activities unfolded. Adam Smith, for example, though acknowledging the existence of more than one factor of production, insisted that the different component parts of price are "measured by the quantity of labour which they can, each of them, purchase or command. Labour measures the value not only of that part of price which resolves itself into labour, but of that which resolves itself into rent, and of that which resolves itself into profit" (Smith 1776/1976, Bk. I, Ch. VI, p. 56). Smith defended the doctrine that outlays on wages determine relative prices, and Ricardo's chapter I, section I is devoted to attacking this doctrine, charging Smith that his measuring rod, the purchasing power of commodity over labour, will not do: Smith had erroneously identified a labour-embodied approach with a labour-commanded one and tried to explain relative prices with a labour-commanded approach. Ricardo's counterargument is that the amount of labour that a product can command in exchange constitutes a poor measure of value. Ricardo's move in this explanatory game has been "in other words, that it is the comparative quantity of commodities which labour will produce, that determines their present or past relative value, and not the comparative quantities of commodities, which are given to the labourer in exchange for his labour" (Ricardo 1817/1951, p. 17). Malthus (1814, p. 12), for his part, was very critical of viewing labour as the measure of value:

> Adam Smith was evidently led into this train of argument, from his habit of considering labour as the standard measure of value, and corn as the measure of labour. But that corn is a very inaccurate measure of labour, the history of our own country will amply demonstrate; where labour, compared with corn, will be found to have experienced very great and striking variations, not only from year to year, but from century to century; and for ten, twenty, and thirty years together. And that neither labour nor any other commodity can be an accurate measure of real value in exchange, is now considered one of the most incontrovertible doctrines of political economy; and, indeed, follows from the very definition of value in exchange.

John Stuart Mill adopts the doctrine that value depends principally on the quantity of labour required to produce goods and argues that prices of commodities produced by labour of different skills are affected by differences in relative wages, but "considering the causes of variations in value, quantity of labour is the thing of chief importance" (Mill

1848/1909, Sec. III. 4.8). Other classical economists offered a series of arguments about and criticisms of the main rule of inference used to explain the value of commodities: the law (or doctrine) that value depends on the quantity of labour.

In this first period of the explanatory game there was a specific cognitive division of labour. It was Adam Smith who provided the big intuitions, David Ricardo who specified more rigorously the rules of inference and John Stuart Mill who extended their scope of applications. The explanations that have been generated while these and other less memorable classical economists were playing the game were the outcome of the complex processes of social interaction between individuals pursuing different theoretical and practical aims. A flow of explanations of the value of commodities was produced and criticized mainly on theoretical terms since in this case no direct experimental evidence was produced. It has mainly been these theoretical criticisms that have gradually led to the change in the rules of inference; and it was the introduction of mathematical models that gradually led to a new set of rules of representation. The scope of the analysis has also been extended to include monopolies and (later) oligopolistic competition. In this specific explanatory game, the change concerned all kinds of rules: the rules of inference, the rules of representation and the rules of scope. The change was initiated by the marginalists. These economists, Walras, Jevons and Menger being the most influential among them, were still playing the same game since the constitutive rules had not changed: the explanandum remained the value of commodities, the metaphysical assumptions remained largely the same and a great part of what was taken for granted by the classical economists was also taken for granted by the marginalists. The main insight had to do with the acknowledgment that value is nothing inherent in goods – "nichts den Gütern Anhaftendes" as Menger pointedly remarked (1871/1968, p. 86) – but a judgement economizing men make about the importance of the goods at their disposal for the maintenance of their lives and well-being – "ein Urtheil, welches die wirtschaftenden

Menschen über die Bedeutung der in ihrer Verfügung befindlichen Güter für die Aufrechthaltung ihres Lebens und ihrer Wohlfahrt fällen" (1871/1968).[2]

The *rules of inference* started changing when economists confronted a series of difficulties in explaining prices because of the lack of a theory of demand. The concept of substitution at the margin provided a more satisfactory explanation of prices of products and factors of production. Classical economists derived the prices of products from the "natural" rates of reward of the three factors of production, which were in turn explained by three separate principles: wages of labour were determined by the long-run costs of producing the means of subsistence; land rentals were determined as a differential surplus over the marginal costs of cultivation; and the rate of profit on

---

[2]   Here is the passage in the German original (1871/1968, p. 86) and the English translation (1976, p. 120f.):

Der Werth ist demnach nichts den Gütern Anhaftendes, keine Eigenschaft derselben, eben so wenig aber auch ein selbstständiges, für sich bestehendes Ding. Derselbe ist ein Urtheil, welches die wirtschaftenden Menschen über die Bedeutung der in ihrer Verfügung befindlichen Güter für die Aufrechthaltung ihres Lebens und ihrer Wohlfahrt fällen, und demnach ausserhalb des Bewusstseins derselben nicht vorhanden Es ist demnach auch durchaus irrig, wenn ein Gut, welches für die wirtschaftenden Subjecte Werth hat, ein "Werth" genannt wird, oder aber die Volkswirthe gar von "Werthen," gleichwie von selbstständigen realen Dingen sprechen, und der Werth solcherart objectivirt wird. Denn das, was objectiv besteht, sind doch immer nur die Dinge, beziehungsweise die Quantitäten derselben, und ihr Werth ist etwas von denselben wesentlich verschiedenes, ein Urtheil nämlich, welches sich die wirtschaftenden Individuen über die Bedeutung, welche die Verfügung über dieselbe für die Aufrechterhaltung ihres Lebens, beziehungsweise ihre Wohlfahrt hat. Es hat aber die Objectivierung des seiner Natur nach durchaus *subjectiven* Güterwerthes gleichfalls sehr viel zur Verwirrung der Grundlagen unserer Wissenschaft beigetragen.

Value is thus nothing inherent in goods, no property of them, nor an independent thing existing by itself. It is a judgment economizing men make about the importance of the goods at their disposal for the maintenance of their lives and well-being. Hence value does not exist outside the consciousness of men. It is, therefore, also quite erroneous to call a good that has value to economizing individuals a "value", or for economists to speak of "values" as of independent real things, and to objectify value in this way. For the entities that exist objectively are always only particular things or quantities of things, and their value is something fundamentally different from the things themselves; it is a judgment made by economizing individuals about the importance their command of the things has for the maintenance of their lives and well-being. Objectification of the value of goods, which is entirely *subjective* in nature, has nevertheless contributed very greatly to confusion about the basic principles of our science.

capital was regarded as a residual. The value of land, of capital and labour was thus explained by different rules. The marginal economists introduced a new rule of inference, the maximization principle, so that distribution became nothing more than the outcome of the application of a more general principle.[3] So, the marginalists achieved greater generality and unification by explaining both factor and product prices with the help of a single principle (Mantzavinos 1999, p. 686).[4]

It is important to note that substitution at the margin, which has been the core novelty in constructing an explanation of the phenomenon of the value of goods along with the maximization principle, does not necessarily presuppose the use of specific rules of representation. None of the leading figures of the Austrian School, that is, Menger, von Wieser or von Böhm-Bawerk, ever used mathematics when formulating and applying the maximization principle. They

---

[3]   Menger has provided a very clear formulation of this principle (1871/1968, p. 158f.):

Dasselbe Princip nun, welches die Menschen in ihrer wirtschaftlichen Thätigkeit überhaupt leidet, *das Bestreben, ihre Bedürfnisse möglichst vollständig zu befriedigen,* dasselbe Princip also, das die Menschen dazu führt, die Nützlichkeiten in der äusseren Natur zu erforschen und ihrer Verfügung zu unterwerfen, dieselbe Sorge nach Verbesserung ihrer wirtschaftlichen Lage, führt nun dieselben auch dazu, die obigen Verhältnisse, wo immer sie vorliegen, auf das Eifrigste zu erforschen und zum Zwecke der besseren Befriedigung ihrer Bedürfnisse auszubeuten, das ist, in unserem Falle zu bewirken, dass jene Güterübertragung, von der wir oben sprachen, auch thatsächlich erfolge. Es ist dies aber die Ursache alle jener Erscheinungen des wirtschaftlichen Lebens, welche wir mit dem Worte *"Tausch"* bezeichnen [...].

In the English translation (1976, p. 179f.):

The same principle that guides men in their economic activity in general, that leads them to investigate the useful things surrounding them in nature and to subject them to their command, and that causes them to be concerned about the betterment of their economic positions, *the effort to satisfy their needs as completely as possible,* leads them also to search most diligently for this relationship wherever they can find it, and to exploit it for the sake of better satisfying their needs. In the situation just described, therefore, the two economizing individuals will make certain that the transfer of goods actually takes place. The effort to satisfy their needs as completely as possible is therefore the cause of all the phenomena of economic life which we designate with the word *"exchange".*

[4]   See also Ott and Winkel (1985, p. 224), Streissler (1989, pp. 128ff.) and Brandt (1993, pp. 277ff.).

were not only innocent of any mathematics but more than that they opposed on methodological grounds the representation of economic phenomena by mathematical means.[5]

The introduction and use of mathematical rules of representation was clearly a distinct novelty, not connected with the introduction of utility theory and subjective theory of value. The use of mathematics for the study of market phenomena has been championed by August Cournot, who was the first to draw a demand curve in the fourth chapter of his *Recherches sur les Principes Mathématiques de la Théorie des Richesses* already published in 1838. He was the first to express the functional dependence between the actual quantities that the consumers purchased annually and the average of annual prices, that is, the empirical relationship between sales and prices, by means of a function (1838, p. 37):

> Admettons donc que le débit ou la demande annuelle D est pour chaque denrée, une fonction particulière F(p) du prix p de cette denrée. Connaître la forme de cette fonction, ce serait connaître ce que nous appelons la loi de la demande ou du débit. Elle dépend évidement du mode d' utilité de la chose, de la nature des services qu'elle peut rendre ou des jouissances qu'elle procure, des habitudes et des mœurs de chaque peuple, de la richesse moyenne et d' échelle suivant laquelle la richesse est répartie.[6]

Mathematical calculations, more specifically maximization under constraints, already used by Cournot, became widespread, and in the

---

[5] See the letter of Carl Menger to Leon Walras from February 1884 reprinted in Jaffé (1965, pp. 2–6).

[6] See the correct observation of Blaug (1997, p. 301f.): "Cournot, in a book that for sheer originality and boldness of conception has no equal in the history of economic theory, was the very first writer to define and to draw a demand function. He took no interest in utility theory but assumed as a matter of course that the market demand curve was negatively inclined: this market demand curve did not express the quantities which the sum of consumers in a market would purchase at different prices, holding constant 'population, and the distribution of wealth, tastes, and the habits of the consuming population', but rather the actual quantities they did purchase annually at an average of annual prices – Cournot's demand curve is an empirical relationship between sales and prices."

end the maximization principle came to be expressed with the aid of these *new rules of representation*. The general rule is to order a series of attainable positions in terms of the respective associated values of a relevant maximand, and then to determine the optimum as the position that assigns the greatest value to the maximand. The maximand can be either utility or physical product or profit, but the formal representational rule remains the same. The application of this formal means of representation has made the explanations more precise, simpler, and offered a unification of diverse economic phenomena.

Johann Heinrich von Thünen's *Der isolierte Staat in Beziehung auf Landwirtschaft und Nationalökonomie*, published in 1826 and extended in a second edition in 1842, made extensive use of mathematics and propagated the utility of the formal tools of mathematics in representing economic laws:

> Aber die Anwendung der Mathematik muß doch da erlaubt werden, wo die Wahrheit ohne sie nicht gefunden werden kann. Hätte man in anderen Fächern des Wissens gegen den mathematischen Kalkül eine solche Abneigung gehabt, wie in der Landwirtschaft und der Nationalökonomie, so wären wir jetzt noch in völliger Unwissenheit über die Gesetze des Himmels; und die Schiffahrt, die durch die Erweiterung der Himmelgründe jetzt alle Weltteile miteinander verbindet, würde sich noch auf die bloße Küstenfahrt beschränken.
>
> *(Thünen 1842, p. 569)*

Von Thünen's work was connected to marginalism since he has consistently applied the principle that all forms of expenditure should be carried to the point at which the product of the last unit equals its cost, that is, in order for the total product to be maximized, resources should be allocated equimarginally. The second volume of his *Isolated State* published in 1850 includes an early example of the use of differential calculus to solve a maximization problem. In this way von Thünen has offered a different and novel kind of representational vehicle on the way towards the explanation of economic phenomena. The connection of marginalism as a new set of *rules of inference*, most prominently the maximization principle, with a new set of *rules of representation*, most prominently the differential calculus, has taken place most influentially in the work of Jevons and Walras. And one should keep in mind that as Blaug (1997, p. 292) correctly points out the notion of reducing social phenomena to mathematical equations was a novelty and "profoundly disturbing to nineteenth century readers."

William Stanley Jevons's *Theory of Political Economy* published in 1871 offers a good example of the interplay of the novel rules of inference and novel rules of representation in the explanation of value as stated in the following quotation (1871, p. 59f.):

> The principles of utility may be illustrated by considering the mode in which we distribute a commodity when it is capable of several uses. There are articles which may be employed for many distinct purposes: thus, barley may be used either to make beer, spirits, bread, or to feed cattle; sugar may be used to eat, or for producing alcohol; timber may be used in construction, or as fuel; iron and other metals may be applied to many different purposes. Imagine, then, a community in the possession of a certain stock of barley; what principles will regulate their mode of consuming it? Or, as we have not yet reached the subject of exchange, imagine an isolated family, or even an individual, possessing an adequate stock, and using some in one way and some in another. The theory of utility gives, theoretically speaking, a complete solution of the question.
>
> Let s be the whole stock of some commodity, and let it be capable of two distinct uses. Then we may represent the two quantities appropriated to these uses by x_1 and y_1 it being a condition that $x_1 + y_1 = s$. The person may be conceived as successively expending small quantities of the commodity. Now it is the inevitable tendency of human nature to choose that course which appears to offer the greatest advantage at the moment. Hence, when the person

remains satisfied with the distribution he has made, it follows that no alteration would yield him more pleasure; which amounts to saying that an increment of commodity would yield exactly as much utility in one use as in another. Let Δu1, Δu2, be the increments of utility, which might arise respectively from consuming an increment of commodity in the two different ways. When the distribution is completed, we ought to have Δu1 = Δu2; or at the limit we have the equation

$$\frac{du_1}{dx} = \frac{du_2}{dy}$$

which is true when x, y are respectively equal to x_1, y_1 We must, in other words, have *the final degrees of utility* in the two uses equal.

The *scope* of the explanatory game has also been extended to include the explanation of prices under monopolistic conditions. The classical economists did address the case of monopoly prices, but in a very unsystematic way since their main rule of inference, that value depends on the amount of labour, could only work under competitive conditions.

Cournot (1838) was the first to offer not only a theory of pure monopoly but also a theory of duopoly and the idea that perfect competition is the limiting case of the spectrum of market forms defined in terms of the number of sellers. But it was Leon Walras who has extended the scope of the new kind of explanations by introducing the concept of general equilibrium and by showing how all product and factor markets in an economy are interdependent. Once one has shown that consumers with given money incomes maximize utility relative to prices ruling in the market so as to obtain the same marginal utility per dollar from every product they purchase; and at the same time producers maximize profits relative to product and factor prices by employing those quantities of the factors of production in order to obtain the same marginal value product per dollar of factor outlays, the question of *scope of this explanation* emerges. Walras in his *Élements d' Économie Politique Pure, ou Théorie de la*

Richesse Sociale (1874) inquires into the question of how the sum of the demand prices of the consumers in a particular product market yields the market demand and analogically how the individual supply prices of the participating firms in the market yield the market supply, and how demand and supply converge on an equilibrium in a single market. Moreover, he extends this analysis to all product and factor markets and shows in a rigorous way, with the use of mathematical rules of representation, what are the conditions for multi-market equilibria to take place and gives the general intuition that all product and factor markets are interdependent a rigorous, formal expression. The procedure that he suggested in all cases of multi-commodity exchange, was to write down the abstract demand and supply equations assuming perfect competition, perfect factor mobility, and perfect price flexibility and then to provide the proof of the existence of a general equilibrium solution for this specific set of simultaneous equations by simply counting the number of equations and the number of unknowns. In case they were equal, he concluded that a general equilibrium solution was at least possible and suggested that a process of quasi-automatic adjustment of price in response to excess demand or supply would take place, *tâtonnement*, that is, groping by trial and error.

This change of the *rules of representation*, the *rules of inference* and the *rules of scope* came about as the outcome of a process of criticism of the explanatory activities of those economists still employing the initial set of rules by an ever-growing number of economists who propagated the advantages of the new rules. Those advantages were mainly conceived in terms of increased consistency of the rules, increased accuracy and simplicity. The unification provided by the extension of the scope of the explanatory game also counted as an additional benefit. The successful propagation of the new rules by a few innovators called more and more imitators to the scene, and so a widespread adoption of the new rules of the explanatory game took place in a gradual learning process. Whether this change of rules had a revolutionary character or a more gradual one is an issue that I will leave undiscussed here. It is only important to stress that under the new set of rules, new kinds of explanatory activities started generating novel explanations of market prices of commodities at different levels.

Of course, this process was neither simple nor straightforward. The remarkable case of Hermann Heinrich Gossen nicely exemplifies that. In his *Entwicklung der Gesetze des menschlichen Verkehrs, und der daraus fließenden Regeln für menschliches Handeln*, published in 1854, he clearly states the law of diminishing marginal utility as well as the law that a consumer will allocate his available income so that the ratio of marginal utility to price is equal across all commodities:

> Die Größe eines und desselben Genusses nimmt, wenn wir mit Bereitung des Genusses ununterbrochen fortfahren, fortwärhend ab, bis zuletzt Sättigung eintritt.
>
> *(Gossen 1854, p. 4f.)*
>
> Der Mensch, dem die Wahl zwischen mehren (sic!) Genüssen frei steht, dessen Zeit aber nicht ausreicht, alle vollaus sich zu bereiten, muß, wie verschieden auch die absolute Größe der einzelnen Genüsse sein mag, um die Summe seines Genusses zum Gräßten zu bringen, bevor er auch nur die größten sich vollaus bereitet, sie alle theilweise bereiten, und zwar in einem solchen Verhältnis, daß die Größe eines jeden Genusses in dem Augenblick, in welchem seine Bereitung abgebrochen wird, bei allen noch die gleiche bleibt.
>
> *(Gossen 1854, p. 12)*

Both these laws flow from the main hypothesis of utility maximization which in fact makes up the opening statement of his opus:

> Der Mensch wünscht sein Leben zu genießen und setzt seinen Lebenszweck darin, seinen Lebensgenuß auf die möglichste Höhe zu steigern.
>
> *(Gossen 1854, p. 1)*

Although Gossen has clearly developed both the rules of inference and the mathematical rules of representation of marginalism and applied them consistently on a broad array of issues, his work has been neglected for decades and had no impact whatsoever in the process of the replacement of the old sets of rules with the new ones. It was only after Jevons and Walras had successfully published and propagated their own versions of marginalism that they discovered the existence of the book of Gossen. It is characteristic how Jevons refers to Gossen in the Preface of the second edition of his *Theory of Political Economy* published in 1879:

> We now come to a truly remarkable discovery in the history of the branch of literature. Some years since my friend Professor Adamson had noticed in one of Kautz's works on Political Economy a brief reference to a book said to contain a theory of pleasure and pain, written by a German author named Hermann Heinrich Gossen [...]
>
> Gossen evidently held the highest possible opinion of the importance of his own theory, for he commences by claiming honours in economic science equal to those of Copernicus in astronomy. [...]
>
> The coincidence, however, between the essential ideas of Gossen's system and my own is so striking, that I desire to state distinctly, in the first place, that I never saw nor so much as heard any hint of the existence of Gossen's book before August 1878, and to explain, in the second place, how it was that I did not do so. My unfortunate want of linguistic power has prevented me, in spite of many attempts, from ever becoming familiar enough with German to read a German book. I once managed to spell out with assistance part of the logical lecture notes of Kant; but that is my sole achievement in German literature. [...]
>
> I cannot claim to be totally indifferent to the rights of priority; and from the year 1862, when my theory was first published in brief outline, I have often pleased myself with the thought that it was at once a novel and important theory. From what I have now stated in this preface it is evident that novelty can no longer be attributed to the leading features of the theory. Much is clearly due to Dupuit, and of the rest a great share must be assigned to Gossen. Regret may easily be swallowed up in satisfaction if I succeed eventually in

making that understood and valued which has been so sadly neglected.

Almost nothing is known to me concerning Gossen; it is uncertain whether he is living or not. On the title-page he describes himself as Königlich preussischem Regierungs – Assessor ausser Dienst, which may be translated 'Royal Prussian Government Assessor, retired'; but the tone of his remarks here and there seems to indicate that he was a disappointed if not an injured man. The reception of his one work can have lent no relief to these feelings; rather it must much have deepened them. The book seems to have contained his one cherished theory; for I can find under the name Gossen no trace of any other publication or scientific memoir whatever. The history of these forgotten works is, indeed, a strange and discouraging one; but the day must come when the eyes of those who cannot see will be opened. Then will due honour be given to all who like Cournot and Gossen have laboured in a thankless field of human knowledge, and have met with the neglect or ridicule they might well have expected. Not indeed that such men do really work for the sake of honour; they bring forth a theory as the tree brings forth its fruit.

Gossen never held a position at a university nor published anything else than this book. He was so bitterly disappointed with the reception of his work that he recalled all the unsold copies from the publisher (who had published it on commission only) and destroyed them. Afflicted with tuberculosis soon afterward, Gossen died in 1858.

6.3   A CASE STUDY FROM MEDICINE

As a further exemplification of an abstract characterization of a game, I will discuss next an explanatory game in the history of cardiology that concerns the functioning of the heart and the circulation of the blood, that is, the way the cardiovascular system works. This is a very interesting case not only because of the obvious importance of the explanation of the way that the human circulatory system functions and the specific role of the heart in it, which is something that has sparked the interest of physicians since antiquity, but mainly because of the quite spectacular fact that a change of the rules of the explanatory game took about fifteen centuries to unfold. The influence of Claudius Galen remained so formidable from the end of the second century CE until the beginning of the sixteenth century CE that Jacobus Sylvius, a leading authority of the sixteenth century, still firmly believed that Galen was infallible, that Galen's *De Usu Partium* was divine, and that any knowledge past Galen was simply impossible (Roth 1892). I will start with the explanatory game that Galen and his contemporaries were playing in the second century and will then show how this game was transformed into the one that William Harvey and his contemporaries came to play in the middle of the seventeenth century.

1. The *constitutive rules* of the explanatory game consisted first (a) in what counted as an explanandum. Since classical antiquity, and certainly at the time of Galen, the blood circulation and the functioning of the heart was an unambiguously delineated phenomenon to be explained. Further (b) the explanatory game took place within a context of background knowledge which remained unquestioned. This bedrock of unquestioned facts and beliefs was omitted from the explanatory game largely in an implicit manner. Finally (c) a series of metaphysical presuppositions were important for this specific game. They had to do with a broadly teleological worldview according to which "Nature does nothing in vain", as Galen characteristically stresses in his *De Usu Pulsuum* (V 155–156),[7] a doctrine to be drawn upon for the explanations of the structure and function of human body. More important still was the postulation of *pneuma*. The idea of a "life-giving spirit" was the most important concrete metaphysical idea that was presupposed and was – as we will shortly see – a constitutive part of the explanations produced. The stages of *pneumatic* elaboration are summarized by Galen in his *De Usu*

---

[7] See the English translation as *On the Function of the Pulse* by Furley and Wilkie (1984, p. 200).

Partium (III 541–542):[8] "From the outside air, *pneuma* is drawn in by the rough arteries and receives its first elaboration in the flesh of lungs, its second in the heart and the arteries, especially those of the retiform plexus, and then a final elaboration in the ventricles of the brain, which completes its transformation into psychic *pneuma*."

2. The *rules of representation* are of special interest mainly due to their uniquely direct character. The description of the phenomenon has largely been obtained by means of dissection. This allowed a direct representation of the heart and the blood circulation via all five senses: sight, hearing, touch, smell and taste. This is probably as clear and direct a depiction of an explanandum phenomenon as possible – and at the same time it is an example that the clear and direct sensual input does not automatically translate into an accurate description: it was only with the use of the microscope by Malpighi in the middle of the seventeenth century that an accurate description of the heart and blood circulation was made possible. Other means of representation, more common indeed, such as the use of natural language and most probably also of diagrams showing the cardiovascular system, were also used.

3. The *rules of inference* used were mainly the rules of logic and those of analogical reasoning. Galen and his contemporaries were not allowed access to human bodies by dissection, though they did see inside humans in surgery and by chance. It was the heart, veins, arteries, livers and the other organs of dead animals that could be directly represented after a dissection and those of living animals that could be directly represented after a vivisection. The rule of inference was then used that the cardiovascular system of animals functioned in an analogous way to the cardiovascular system of humans. The employment of this analogy, as the main rule of inference, was, as is evident, of a dramatic importance for the explanations produced.

---

[8] The text is cited according to the edition of Helmreich, G. *Galeni. De Usu Partium Libri XVII*, Leipzig: Teubner, 1907–1909, and the quotation is to be found in i 393, 23–394, 6.

4. The *scope* of the explanations provided was straightforward and concerned the cardiovascular system, that is, the heart, the veins and the arteries. However, the explanatory game dealing with the functioning of the heart and of blood circulation was evidently nested into the explanatory game dealing with human nutrition, since the main function of blood was to provide the distribution of food to all parts of the body. This in turn was nested in the explanatory game for human physiology, and this was made possible by the rules of scope, since the veins and arteries connect all parts of the human body with the heart.

In a recent paper in the *Journal of Thrombosis and Haemostasis*, Aird (2011, p. 120f.) summarizes how the functioning of the cardio-vascular system is explained by Galen. I quote this at length since all rules are at work here (p. 182):

> *Veins contain blood.* According to Galen, the liver is the source of all veins and principle instrument of sanguification. In the stomach, food is concocted into chyle, which is then delivered to the small intestine and absorbed into veins. The chyle is carried in the portal vein to the liver, where nutriment becomes actual blood, which is charged with natural spirits. Blood is purified in the liver and then enters the hepatic vein through invisible connections between branches of the portal and hepatic veins. The blood moves from the hepatic vein to the inferior vena cava, which through its branches supplies all the parts of the body above and below the liver. In other words, blood moves centrifugally from the center (the liver) to the periphery. This is an open-ended system designed to provide one-time distribution of food. Each part of the body attracts and retains only enough blood for its immediate requirements. Blood that is assimilated into tissue is ultimately lost through invisible emanation. The parts receive fresh supplies from the liver as needed. As such, movement of blood was subsumed under the *theory of nutrition* according to which each body attracts, retains, and assimilates food, and expels its superfluities.

A portion of blood nourishes the lung via the right ventricle.
A small amount of blood entering the vena cava is diverted to the
right auricle, which is considered an outgrowth of the caval system.
From the right auricle blood enters the right ventricle. Dilatation of
the right ventricle draws in blood from the vena cava. The right
ventricle further elaborates and attenuates the blood, rendering it
fine and thin. Some of this refined blood enters the pulmonary
artery. Blood in the pulmonary artery nourishes the lung. A small
portion of the blood (the thinner part) in the pulmonary artery is
squeezed through invisible anastomoses into the pulmonary veins,
from which it too is absorbed by the lungs, providing them with
vital spirits. Finally, some blood in the right ventricle passes into
the left ventricle through invisible pores in the interventricular
septum.

The heart intrinsically pulsates. Galen recognized that both
ventricles pulsate even when their nerves are severed or the heart
is removed from the thorax. Thus the power of pulsation has its
origin in the heart itself. [...]

Respiration cools the innate heat and yields vital spirits. [...]

Arteries contain air and blood. [...]

The whole body breathes in and out. [...]

All is in all. None of the parts of the body is absolutely pure.
Everything shares in everything else. Thus, while arteries are
primarily instruments of the *pneuma*, they have their share of
thin, pure spirituous blood. Veins are primarily instruments of the
blood or other nutriment, but contain a little mistlike air. All over
the body, arteries and veins communicate with one another by
common openings and exchange of blood and *pneuma* occurs
through certain invisible and extremely narrow passages or
inosculations. Through these junctions, the arteries draw from the
veins, when they expand, and squeeze into them when contracting.
Thus, the movement of blood and air is neither directional nor
rapid. Rather, the contents of blood vessels move slowly, hither
and thither.

Explanations of the functioning of the cardiovascular system of the type summarized here, though mainly crystallized and handed down over the centuries as contained in the impressive Galenic corpus (which accounts for about 10 per cent of what we possess of Greek prior to 350 CE),[9] were produced according to the standards prevailing at that time and *in a process of social interaction unfolding within the given institutional rules*. A feature of medicine as it was practised at that time was that several doctors were often summoned to the patient's bedside (Mattern 2008, pp. 71ff.), where they offered competing diagnoses and prognoses, and it was up to the patient himself or his representatives to choose among them (Hankinson 2008, p. 8). The academic reputation was partly earned in public disputes where rhetorical skills, among others, were essential for success;[10] in *De Optimo Medico Cognoscendo* (105, 4–15, Iskandar 1988), in his notorious or even perhaps infamous autobiographical manner,[11] Galen himself describes the incident that led to his appointment as the physician of the gladorial school in Pergamum in 157 CE for four years:

> Once I attended a public gathering where men had met to test the knowledge of physicians. I performed many anatomical demonstrations before the spectators: I made an incision in the

[9]   According to the estimation of Nutton (2004, p. 390, n. 22) taking an ancient book-roll as the equivalent of 30 printed pages, he wrote/dictated 2 or 3 pages a day over a working life of some 60 years, on top of his other activities. Kühn's standard edition (1821–1833), which contains the Greek text with subjoined Latin translation, is about 16,000 pages long; and discoveries since then of versions of some of his lost works in Latin or Arabic have added further to the length of his opus.

[10]   As Hankinson notes (2008, p. 11): "Public demonstration, or demonstration before an influential invited audience, of either scientific or argumentative skill [...] was a standard feature of the intellectual life of the times [it also served as a rather cruder form of entertainment, at any rate in the case of the vivisectional demonstrations]."

[11]   The autobiographical works of Galen seem to be of a self-promoting character even verging sometimes on autohagiography, as Hankinson (2008, p. 7) observes. However, in her biography of the *Prince of Medicine* Winter (2013, p. 4) draws attention to the conditions prevailing in his time: "In the modern western world he might be diagnosed with a personality disorder, once megalomania, today narcissism. He also epitomizes the much-maligned 'type A' personality. All of this, however, was typical of his time, place, and social stratum, and Galen was not more competitive, hostile or self-aggrandizing than his peers."

abdomen of an ape and exposed its intestines: then I called upon
the physicians who were present to replace them back (in
position) and to make the necessary abdominal sutures – but
none of them dared to do this. We ourselves then treated the
ape displaying our skill, manual training, and dexterity.
Furthermore, we deliberately severed many large veins, thus
allowing the blood to run freely, and called upon the Elders of
the physicians to provide treatment, but they had nothing to
offer. We then provided treatment, making it clear to the
intellectuals who were present that (physicians) who possess
skills like mine should be in charge of the wounded. That man
was delighted when he put me in charge of the wounded – and
was the first to entrust me with their care.

But of course neither in the incident narrated earlier nor in the theoretical writings of Galen were the rhetorical skills the fundamental
factor that led to explanatory success. The explanatory game unfolds
in constant interaction with Nature and with the other players, so that
the explanations produced must successfully account for the natural
phenomena – one way or another – in order to be accepted by the other
players over time. Mere rhetoric and charlatanism are usually not
sufficient, and they were certainly not sufficient in the case at hand.
Galen's explanations managed to acquire a prominent status among
his peers (Staden 1995, pp. 51ff.) independently of the fact that at some
periods of his life, notably in Rome, he was making public demonstra-
tions on a daily basis (*De Venae Sectione adversus Erasistrateos
Romae Degentes* XI 194).[12] This was because they were based on
solid experiential evidence and the *correct use of logical and other
inferential rules*.[13]

---

[12]   See the English translation by Brain (1986).

[13]   See Hankinson (2008, p. 24): "If nothing else, Galen's vast literary output, over a period
    of perhaps seventy years, when he was constantly engaged in other activities, is
    testament to his prodigious energy and industry; while his undoubted rhetorical
    excessiveness, so grating to many modern ears, is none the less characteristic of its
    times. There is no doubt that Galen's texts are rhetorical; no doubt that he is the hero
    of his own story; and no doubt that Galen sometimes misrepresents the positions of

On one occasion Galen challenged the Erasistrateans, who followed the view that arteries contain air or pneuma alone, to show him an artery that is empty of blood (*De Anatomicis Administrationibus* VII 10: II 619, pp. 16ff.). He proved that all arteries in the body contain a portion of blood (and so that the position of Erasistratus was not tenable) by ligating an artery in two places, slicing open the intervening segment, and finding blood, but no air. However, the rule of inference that he subsequently used was clearly erroneous; addressing the question of how blood gets from the arteries to the veins he suggested that blood permeates from pulmonary arteries to pulmonary veins through invisible channels. The resulting blood in the pulmonary veins does not reach the left ventricle, but it is rather used as nourishment by the lungs; hence there is no pulmonary circuit. Instead, his main inference was that blood in the left ventricle, and so the systemic arteries, is derived directly from the right ventricle, through invisible pores in the interventricular septum (*De Usu Partium* III 497, = i 362, 19–123 Helmreich; more cautiously *De Anatomicis Administrationibus* II 623, 2–3).[14]

Although Galen did recognize the unidirectional characteristics of the cardiac valves, he was not able to explain the circulation through the heart since he adhered to the view that blood was subject to an ebb-and-flow motion. This remained the main, more general, *inferential*

his opponents in order to sharpen his critique and to emphasize his differences from them. But rhetorical extravagance does not imply falsehood, as some apparently suppose; nor is exaggeration invariably a cardinal sin. Galen saw himself, no doubt in self-aggrandizing terms, as a man on a heroic mission to rescue medicine, and science in general, from their degenerate decrepitude. Desperate times called for desperate measures. And if he was often mistaken, and in general unjustifiably over-confident of the truth of his position and the security of his first principles, he was not incapable of changing his mind, and of learning from his errors, when he cared to admit to them."

[14] Key, Keys and Callahan make the following observation in an article in *The American Journal of Cardiology* (1979, p. 1028): "Galen subsequently abandoned the idea that a significant amount of blood passed through pores in the interventricular septum, although this was accepted for many centuries and was a specific point singled out by Harvey for attack." Acierno (1994, p. 184) mentions this view and seems to adopt it. However, none provides any textual or other evidence that this was indeed the case.

rule. The notion of the pump was not available at that time, so that *drawing such an analogy* and inferring that also the heart functions as a pump was *not among the possible options*. Galen inferred instead that the forward flow of blood from the right ventricle through the lungs was due to the rhythmic respiratory movements of the thorax (Acierno 1994, p. 217). The explanations of the cardiovascular system remained inaccurate, thus, due to the *metaphysical rules* employed that involved pneuma as an ingredient in the blood circulation, due to the *rather limited rules of representation based solely on anatomical means* – dissection of dead animals and vivisection of living ones – and due to the inferential rules employed which, with the exception of the logical ones that secured the validity of argumentation, did not provide accuracy. (Figure 4 contains a *modern sketch* of the circulatory system as conceived in ancient times.)

Of course, there have been controversies with respect to all kinds of rules. In his work *De Sectis ad eos qui Introdunctur*[15] Galen gives a crude classification of the competing schools of medicine, which he categorizes as the "Dogmatists", the "Empiricists" and the "Methodists".[16] What seems to be of greatest importance in the explanatory game concerning the cardiovascular system are the controversies with respect to anatomy upon which I will briefly focus here.[17]

---

[15] See the English translation as *On Sects for Beginners* by Michael Frede (1985).

[16] As Lloyd (2008, p. 41) observes: "To put the matter in the crudest possible terms, this is because the Dogmatists pay insufficient attention to experience, the Empiricists under-estimate the role of theory and argument and the Methodists abandoned pretty well the whole of the framework within which, traditionally, elite Greek medicine has been practiced."

[17] Human dissection and vivisection began and ended with Herophilus and Erasistratus in the third century BCE. Again, the main source of our knowledge on the prevailing opinions remains the Galenic opus, so that the accuracy of the summarized opinions cannot be taken for granted, Galen usually engaging in them in order to show their errors. Alexandria seems to have been the place where anatomical knowledge was developed most – this is where Galen attended lectures by Pelops, who "wrote some very valuable books, but after his death all were destroyed by fire before anyone had copied them" (*De Anatomicis Administrationibus*, II 217–218, 225, = *De Anatomicis Administrationibus* XIV I:184 DLT). Most notable anatomists were Quintus and Marinus, and Galen describes the latter as having "accumulated no small experience in dissections, and it was he himself who had set his hand to and had observed

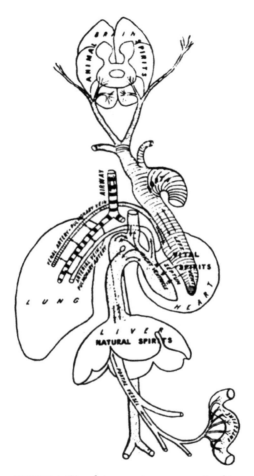

FIGURE 4 Circulatory system as conceived in ancient times.
Courtesy of the US National Library of Medicine

everything that he explained in his writings [...] although now and then we may discover him in error" (*De Anatomicis Administrationibus* XIV I:184–185 DLT. Cf *De Libris Propriis*, XIX 25, =SM 2, 104, 12–13). A man called Martialius seems to have been a main adversary of Galen. He "declared the superiority of Erasistratus in all areas of the art and especially in anatomy", something that provoked Galen to compose the six books on *Hippocrates' Anatomy* and the three on *Erasistratus' Anatomy* and also to use a forum of a public lecture to express his hostility towards him (*De Libris Propriis* 1: XIX 13, 11ff., = SM 2, 94, 26–95, 2).

Empiricists denied the epistemological merit of dissection. According to Celsus, the argument of the Empiricist is that such knowledge is not necessary since what is obtained from vivisection and dissection is not true understanding of the body under normal conditions, given that the very act of dissection produces significant changes in the appearance of the structures under investigation:

> For when the body had been laid open, colour, smoothness, softness, hardness and all similars would not be such as they were when the body was untouched; because bodies, even when uninjured yet often change in appearance, they note, from fear, pain, want of food, indigestion, weariness and a thousand other mediocre affections; it is much more likely that the more internal parts, which are fast softer, and to which the very light is something novel, should under the most severe of woundings, in fact mangling, undergo changes. Nor is anything more foolish, they say, than to suppose that whatever the condition of the part of a man's body in life, it will also be the same when he is dying, nay, when he is already dead [...] That for these reasons, since most things are altered in the dead, some hold that even the dissection of the dead is unnecessary; although not cruel, it is none the less nasty; but all that is possible to come to know in the living, the actual treatment exhibits.
>
> *(Celsus*, On Medicine, Prooemium *40–44)*[18]

The Methodists also saw no need for anatomical research, since Methodist medical epistemology employed reason not in order to search for hidden causes, but for acquiring information about the body which is supposed to be obvious to any thinking person (Frede 1987, p. 265f.).[19] And the dogmatists, too, denied the epistemological merit of dissection.

---

[18]   The translation is from the Loeb Classical Library edition (1935).
[19]   As Rocca observes (2008, p. 246): "Not all doctors, then, performed anatomy to Galen's standards, much less deemed such extensive knowledge relevant for daily practice. Their arguments could not always easily be dismissed and Galen's awesome anatomical erudition should not blind us to this fact."

Controversies with respect to the *rules of inference* to be used in order to explain the cardiovascular system (and the function of all the other organs) were also prevalent. Although the physiology of the heart (and of the whole organism) has affinities with anatomy, it is a distinct field replete with diverse opinions. In our framework *it comprises the set of the inferential rules necessary to provide explanations.*[20]

By 600 CE at the latest Galen's explanatory practices had turned into Galenism: one individual's opinions had become an intellectual system that guided all medical learning. No less authority in ancient medicine than Vivian Nutton (2008, p. 363) observes:

> The Erasistrateans, Pneumatists, Empiricists and Methodists, familiar in Galen's own day, had disappeared before the early sixth century, and, save for the Methodists, perhaps long before then. The lively and wide-ranging medical debates of the Early Roman Empire were replaced by discussions about the proper interpretation of this or that passage in Galen. Galenism had triumphed – arguably to the detriment of Galen. His empiricism, his observational genius and his willingness to think on his feet found little place in Galenism, for its central texts were those that emphasized his conclusions rather than the means by which he had reached them. Anatomical dissection for the purpose of investigation, so much stressed by Galen, seems to have vanished

---

[20] Armelle Debru (2008, p. 264) summarizes the situation in which Galen found himself: "[I]n contrast with anatomy, nothing indicates unequivocally what the role is of certain organs, such as the liver and kidneys, or how functions as complex as digestion and generation come to realize themselves, or above all what further function some of them, such as respiration, might fulfil. Moreover the field was already full of his predecessors' speculations; thus providing an explanation involves refuting other people's opinions as well as defending his own positions and giving the most convincing possible demonstration of them. One needs to show how the thing comes about, what its cause is, and what it is for (this, of course, had already been Aristotle's method). None of this could be done without a rational method, one founded on a mastery of the theory of demonstration, something which required training in both philosophy and logic, and which was the only way of arriving at the truth, such as that which Galen had developed in his great lost work *On Demonstration*."

almost entirely, although both the Byzantines and the Arabs were extremely proficient in surgery.

Galenism then prevailed until the Renaissance, the main reason being

> [i]ts compatibility with the monotheism of the Christians, Muslims and Jews and with the predominant Aristotelianism that explained the natural world [. . .]. Besides, the very fact of its longevity and of the survival and recovery of many patients treated according to Galenic principles reassured its adherents of its efficacy, even in the face of disastrous epidemics such as the Black Death.
>
> *(Nutton 2008, p. 365)*

Human dissection had largely disappeared, to reappear only in the latter part of the Middle Ages when it became an important teaching tool at the University of Bologna sometime in the fourteenth century and finally an integral part of the curriculum by Mondino around 1315 (Siraisi 1981, pp. 110–117). Mondino would lecture in front of his students while a prosector would dissect the cadaver, though it might indeed have been the case that he also performed public human dissections himself – for the first time since Herophilus and Erasistratus. *It was with the aid of evidence in the form of anatomical observation* that Mondino was able to describe quite accurately the heart, especially the details regarding the valves (Acierno 1994, p. 14). This did not lead him to challenge Galen, but has certainly paved the way for criticism by *enabling in principle* the confrontation of Galen's writings with anatomical observations. This was a very small but remarkable step for the times, since the authority of Galen was so formidable that for more than a millennium the dictum was: If there proved to be no holes in the septum, it clearly followed that nature must have undergone changes since Galen (Snellen 1984, p. 21).[21] Henri de Mondeville who held three chairs in Montpellier

---

[21]   As late as 1649, Riolan, one of the followers of Galenism, said that if findings in subsequent dissections differed from the observations of Galen, then nature must have changed. See Key, Keys and Callahan (1979, p. 1028), with reference to Willius and Dry (1948, p. 16).

(in anatomy, surgery and medicine) at the same period as Mondino, that is, at the beginning of the fourteenth century, was able to provide a meticulous description of the heart with the help of anatomy and expressed his *criticism* of Galen's supposed infallibility, declaring that God did not exhaust all his creative power in making Galen.

These first signs of criticism became more and more common in the Renaissance. It was a *novel kind of institutional framework that had gradually emerged* that ultimately enabled the *institutionaliza-tion of criticism* by medical scholars, which in turn led to an incremental change of explanatory rules and the definite abandonment of the explanations designed on the basis of the doctrines of Galenism. As I will discuss in detail in Chapter 9 (of which the present discussion is a good preamble and exemplification), the institutional framework within which explanatory activities unfold consists of informal and formal institutions. The *informal institutions* encapsulate the critical attitude which having started in Ancient Greece was weakened through the centuries, somehow revitalized with the Condemnation of 1277 and revived again during the Scientific Revolution. The incremental spread of an anti-dogmatic attitude during the Renaissance is a complex phenomenon of which I cannot give an account here. It is important only to note that the invention of the printing press in the 1450s provided the possibility of the rapid dissemination of medical knowledge, which along with the increasing, though very slow, legitimization of the questioning of authorities decisively helped human creativity in the domain of medicine to flourish – and more narrowly also in the domain of the explanations of the cardiovascular system that concern us here. The *formal part of the institutional framework* which had to do with the prevailing political institutions was also conducive to allowing criticism and to the unimpaired unfolding of the inquisitive spirit of medical scholars. This was the case in the Republic of Venice, of which Padua has been the university since 1405, which largely guaranteed civil and religious freedom and tolerance. Venice was the most anti-clerical state in Europe; *the organizational structure* of the University of Padua seems to have been quite decisive not only for scientific discoveries in medicine but also for other disciplines during the Scientific Revolution, since also Copernicus and Galileo were at this university during their most productive periods. Thus, "[...] there were developments in that university which would justify the view that in so far as any single place could claim the honour of being the seat of the scientific revolution, the distinction must belong to Padua" (Butterfield 1957, p. 59).

The University of Padua provided the appropriate organizational structure for the development of modern medicine, becoming the centre in or around which scholars providing the explanations of the blood circulation were active. Although it was nominally a Catholic university, it was different than Bologna, Pisa and Rome in that a profession of faith was not required from the students, something which allowed the attendance and graduation of Protestant and Jewish foreign scholars who formed the vast majority of the students (Thiene 1997, p. 81).[22] A sign of the meritocracy actively promoted was the rule that no patrician of Venice or Padua was permitted to hold a chair

---

[22] The *ingenious organizational rule* introduced and followed by the university was sufficient to secure the academic freedom even after the coming of the Reformation and the religious division of Europe in the sixteenth century when universities in Catholic areas such as Italy found it nearly impossible to accept Protestant students (O'Malley 1970, p. 6): "It will be readily understood that sound Venetian business sense was unwilling to permit the loss of income derived annually from the attendance of several thousand Protestants at the university of Padua. Nor, as a matter of pride, did the Republic look forward with pleasure to any consequent decline in the university's prestige. This problem became even more acute when by a papal bull of 1564 Pope Pius IV required that those taking degrees at Catholic universities swear to their profession of the Catholic faith, and at Padua such maladroit papal action created further problems since the degrees were granted in the presence of and nominally by the bishop.

Now it happened that there existed within the German Empire – but theoretically universal in character – the curious title of Count Palatine, an utterly meaningless but not-religious rank which successive emperors had bestowed liberally because it cost them nothing; and among the Count Palatine's perquisites was the declared but hitherto ignored and normally unenforceable right of conferring academic degrees. This became the first solution of the university's dilemma, and in consequence the English, Dutch, German and Swiss Protestant students were able to continue their studies as usual. Count Palatine Sigismund de Capitibus Listae has no other claim to remembrance except that he conferred the degree of Doctor of Medicine upon William Harvey."

in the university – nepotism was, thus, prevented.[23] Every faculty member was allowed full freedom to teach as he desired – there was, however, one specific *organizational rule* that he had to follow: not to repeat his course of lectures in successive years (O'Malley 1970, p. 4).

The following of this rule by the Belgian professor Andreas Vesalius led to the reorganization of the presentations of his anatomical lectures to the effect that what used to be three distinct functions during the anatomical lesson fulfilled by three distinct persons, that is, usually an extraordinary professor of medicine *reading the relevant treatise of anatomy*, the *actual dissection* itself being carried out by the surgeons who were just technicians and the ordinary professor

---

[23]   O'Malley, who provides this piece of information included in the text, gives an excellent overview of the intricate way that the organizational structure has worked, itself embedded in the framework of informal and formal institutions of the time (1970, p. 3f.): "The fact that such developments occurred at Padua, promoting the fame of the medical faculty and increasing its attractiveness for foreign students, including the English, was by no means fortuitous.

Like all Italian universities Padua had begun as an institution controlled by the students, but unlike more recent efforts toward the same goal, student control of Padua had as justification the fact that the students paid all the expenses of the university, that is, the lecturer's fees, for which in return they expected and demanded a proper educational dividend. Control passed out of the hands of the students when those expenses were assumed by the [...] Venetian Government [...] which was shrewd enough to allow the continuance of a student rector of the university, chosen by the students but draped in pomp little real significance in so far as concerned essential educational activities. [...]

Having gained firm control, the Venetian government, to its credit, was genuinely interested in the university's development and in time became immensely proud of the result. [I]n 1517 Venice took upon itself the entire financial burden as well as the immediate direction of the university, hitherto partly borne by the town of Padua. It was at this time that the Paduan magistrates who had formerly controlled the university were replaced by three Venetian overseers known as *Riformatori dello Studio di Padova*. [...] Such then was the new order of administration, characterized by the efficient but enlightened control of Venice which applied certain business and commercial practices to the university with no little success. Both students and professors were subject to the *Riformatori*, under whose direction a number of new regulations were put into effect. For example, members of the faculty were elected for regular terms of up to four years, and only after a longer period of successive reappointments might an exceptional professor, such as Fabrizi d'Acquapendente, be considered a good long-term risk, that is, be elected for life. To prevent pressures from being applied unduly by influential families, and so possibly endangering the high academic quality of instruction, with the exception of individual cases determined individually and solely upon merit, no patrician of Venice or Padua was permitted to hold a chair in the university."

showing the relevant organs, were united in one. Vesalius was the first who combined the role of the lector, sector and ostensor, so that space for a considerable variation in the teaching of what was supposed to be the core discipline of medicine, that is, anatomy, was made possible. This in turn must have been one crucial factor that enabled Vesalius to present an alternative to Galenism in his opus magnum *De Humani Corporis Fabrica Libri Septem* (On the Fabric of the Human Body), which he published in 1543 (the same year as the publication of Copernicus's great work *De Revolutionibus Orbium Coelestium*).

This work *did not provide a novel explanation* of the blood circulation and the functioning of the heart. But Vesalius was the first to *challenge* the existence of pores in the interventicular septum, *one of the main doctrines of Galen,* though as he later confessed himself, he did it very diffidently in the first edition of his work, since he deliberately accommodated his results to conform largely with Galen (Butterfield 1957, p. 56). In the first edition he questioned hesitantly the existence of interventricular septal pores, but he admits of them nevertheless:

> The septum between the ventricles is, as I said, formed from the thickest heart substance and has numerous pits on both sides, this being the main reason for the unevenness of its surfaces that face the ventricles. So far at least as can be determined by the senses, none of these pits goes through from the right ventricle to the left; and we are therefore compelled to marvel at the Creators's clever device by which blood oozes from the right ventricle to the left through invisible channels.
>
> (On the Fabric of the Human Body, *Book VI, Chapter VI, p. 78 [p. 589
> in the original Latin text])*

In the second edition of 1555 he rejects the 1,400-years-old doctrine of Galen more categorically: "Not long ago I would not have dared to turn even a hair's breadth from Galen. But it seems to me that the septum of the heart is as thick, dense and compact as the rest of the heart. I do not see, therefore, how even the smallest particle can be transferred from the right to the left ventricle through the septum."

The main novelty introduced by the *Fabrica* concerns the *rules of representation* that changed dramatically under the influence of this book and largely replaced the rules of representation that were followed for centuries, changing, thus, considerably the way that the explanatory game was played. Vesalius collaborated with Jan Stefan van Calcar from the workshop of Titian towards the construction of woodblock engravings of the drawings contained in the Fabrica – these anatomical drawings and illustrations were the outcome of a productive synthesis of art and science, and reflected a naturalism in the depiction of the human anatomy which was radically different than the conventional medieval drawings. Vesalius supervised personally the actual printing in Basel. With the use of the printing technology, the drawings and diagrams could be copied and multiplied with accuracy and thus the representation of the explanandum phenomenon became more precise and more easily accessible. The direct representation by means of the senses in the process of a dissection was replaced by the printed representation of the heart and the other organs by means of drawings and other illustrations. In our context, the drawing published in his *Tabulae Anatomicae Sex* (plate 3) already in 1538 is of special interest (Figure 5). In the legend to this drawing, Vesalius describes the left atrium (Q) as follows: "arteria venalis in sinistrum sinum aerem ex pulmonibus deferens" (the pulmonary vein carries air from the lungs to the left atrium).

The production of explanations of the functioning of the heart was immensely facilitated, thus, since a standardized and more accurate representation was made possible. The *communication among medical scientists* was made much easier and the conditions for a speedier unfolding of the *explanatory process* were created. In order to understand the importance of such a development in the explanatory game, one needs only juxtapose the anatomical representations of the *Fabrica* and its influence as a canonical anatomical textbook for a very long period of time (its latest printing being in

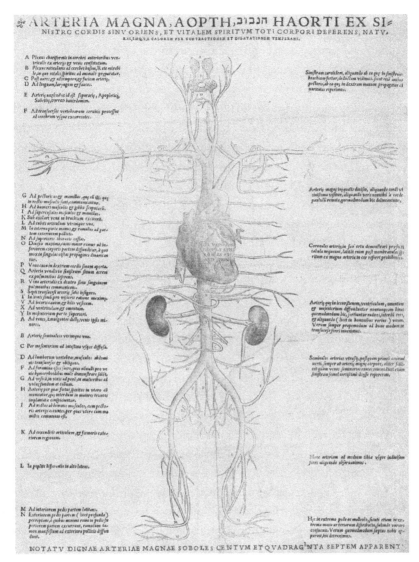

FIGURE 5 Andreas Vesalius: *Tabulae anatomicae sex*, broadside 3, 1538.
Courtesy of University of Glasgow Library, Special Collections

1782) with the drawings of the heart by Leonardo da Vinci, which antedate those of Vesalius by some 30 years. Leonardo created drawings of a two-chambered heart and of the pores in the centrum (Figure 6); he held, however, the correct view that air did not pass from the lungs into the heart and he portrayed in an accurate manner the valves, muscle and coronary vasculature of the heart. But his work was never published during his lifetime and, thus, had no influence whatsoever on the unfolding of the explanatory process. (The collection of his drawings was found in the Windsor collection in the eighteenth century and was first published between 1898 and 1916).

Realdo Colombo, the assistant and later successor of Vesalius as professor of anatomy for two years (1543–1545) in Padua,[24] though remaining within the Galean framework, was able to provide an explanation of what is called the smaller circulation, that is, the passage of the blood from the right side of the heart into the lungs and from there into the left ventricle of the heart. In his single book, *De Re Anatomica* (1559), he noted that the pulmonary artery carried blood to the lungs and the pulmonary vein returned blood mixed with air to the left side. Based on the examination of 1,000 cadavers, according to his statement, and on a series of inferences, he attained an explanation of the smaller circulation, while the description of the general circulation of the blood he offered still aligned with the Galenic doctrine. The *first inference* that he drew was from the observation of the pulmonary veins of cadavers as well as of living animals: the fact that they were full of blood would not have been possible if this vein had to carry only air and fumes. The *second, more important inference* drawn was then that blood must be carried through the pulmonary artery to the lungs and once there, after having been mixed with air, return to the left ventricle of the heart (Colombo 1559, Book 7, transl. from Coppola 1957, p. 62):

> Between these ventricles there is a septum through which almost everyone believes there opens a pathway for the blood from the right

[24] For Colombo's relation with Vesalius see Eknoyan and De Santo (1997, p. 263f).

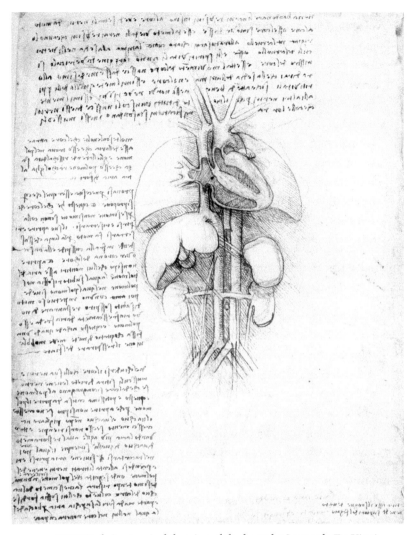

FIGURE 6 Anatomical drawing of the heart by Leonardo Da Vinci
(between 1490 and 1510). Notice the two-chambered heart and the pores
in the septum.
Royal Collection Trust / © Her Majesty Queen Elizabeth II 2016

ventricle to the left, and that the blood is rendered thin so that this may be done more easily for the generation of vital spirits. But they are in great error, for the blood is carried through the pulmonary artery to the lung and is there attenuated; then it is carried, along with air, through the pulmonary vein to the left ventricle of the heart. Hitherto no one has noticed this or left it in writing, and it especially should be observed by all. Wherefore I cannot wonder enough that anatomists have not observed a matter so clear and of such importance. For them it suffices that Galen said so. There are some in our time who swear by the opinions of Galen and assert that he should be taken as gospel and that there is nothing untrue in his writings.

Andrea Cesalpino was a student of Colombo who also provided an explanation of the pulmonary circulation (*Questionum medicarum*, Liber II, Quaestio XVII, 1593, Engl. transl. Prioreschi 2004, p. 384):

> [...] the following must be known: the cavities of the heart are so arranged by nature that from the vena cava the blood flows into the right ventricle of the heart and from there an outlet leads to the lungs. Another passage leads from the lungs to the left ventricle of the heart and there an opening leads to the aorta artery. Certain membranes are placed at the openings of the vessels to preclude the return of the blood so that there is a continuous motion [of blood] from the vena cava through the heart and the lungs to the aorta artery [...].

He not only underlines the continuous flow of blood from the veins to the arteries through the heart but also uses the term "circulation" in the following passage (*Praxis Universae Artis Medicae* 1606, Liber VI, Cap. IX, pp. 469–470, Engl. transl. Prioreschi 2004, p. 384):

> The hot blood is brought to the lungs from the right ventricle by the artery that Galen calls the arterial vein [i.e., pulmonary artery], then it is returned to the heart by the vein going to the left ventricle, which Galen calls the venal artery [i.e., pulmonary vein]; during its passing [through the lungs] the blood is tempered by the cold air

inspired into the bronchi, which are near the veins and arteries, and a sort of circulation of the blood is achieved [...].

A quite *crucial inference* is taking place here, since circulation is explicitly introduced, probably for the first time.

However, there is no explanation of the way the blood was transferred from the arteries to the veins at the periphery. But such an explanatory attempt is already contained in his *Peripateticarum Questionum* (1571, Liber V, Quaestio IV, fol. 111, E–F, Engl. transl. Prioreschi 2004, p. 384) where the "circulation of the blood" is explicitly mentioned:

> Therefore the lung receives the hot blood from the right ventricle of the heart through the vein that is like an artery [i.e., the pulmonary artery] and returns it through the anastomoses with the venal artery [i.e., the pulmonary vein], which goes to the left ventricle. In the meantime it tempers it [i.e., the blood] with cold air brought in through the bronchi (which run near [the branches of] the pulmonary vein) by contact and not by means of openings as Galen thought. The evidence of the circulation of the blood from the right to the left ventricle of the heart through the lungs is confirmed by dissection.

According to Cesalpino "anastomoses" are small openings or *oscula* (i.e., small mouths) through which veins and arteries communicate *(Questionum medicarum*, Liber II, Quaestio XVII, 1593).

It is clear that Cesalpino presented a *set of novel inferences* that addressed not only the pulmonary circulation, but also the circulatory circuit in general. (The debate among generations of medical historians whether Cesalpino or Harvey should be credited with the "discovery of blood circulation" comes under a new light, if one follows my approach of an explanatory game. Explanations are constantly produced and criticized and they are the outcomes of a historical process during which the rules within which the games unfold are constantly changing. The rules of inference that Cesalpino used to draw his conclusions and provide his explanations were not qualitatively different than the ones used before him: these were general inferential rules leading from observations reached by the use of his senses to more general conclusions. This was not the case with Harvey, who, as we will see shortly, was the first to use an arithmetic example, and thus a quasi-mathematical rule of inference for the first time in the provision of his explanation of the blood circulation. The *communication* of the explanation of Harvey was provided in a more systematic and synthetic fashion by means of a short monograph focused on the issue rather within the framework of general treatises and was thus more *persuasive* in the sense that more of his peers came to accept his explanation.)

In *De Venarum Ostiolis* published in 1603 Hieronymus Fabricius ab Acquapendente gave a detailed description of the valves within the veins. The focus on them was important because the inference was drawn that the operation of the valves was to check only the outward passage of blood from the heart and through the veins to the diverse parts of the body. In other words, remaining within the general Galean framework (according to which blood ebbs and flows in the arteries distributing the vital spirit and blood moves similarly in the veins nourishing the body but generally in an outward direction from the heart), Fabricius designated that the valves function to slow the centrifugal flow of blood to the periphery and so to prevent an excessive outward movement of blood – due mainly to peripheral attraction – to the lower part of the limbs at the expense of under-nourishing the upper parts. Since the larger inferential framework remained the Galean one, the discovery of the valves did not lead Fabricius to draw the inference that the normal passage was in the opposite direction, that is, inwards, towards the heart, rather than outwards and, thus, that the blood in the veins was in its return journey (*De Venarum Ostiolis* 1603, facsimile ed., Engl. translation K.J. Franklin 1933, p. 47):

> Valves of veins is the name I give to some extremely delicate little membranes in the lumen of veins. They occur at intervals, singly or

in pairs, especially in the limb veins. They open upwards in the direction of the main venous trunk, and are closed below, while, viewed from the outside, they resemble the swellings in the stem and small branches of plants. My theory is that Nature has formed them to delay the blood to some extent, and to prevent the whole mass of it flooding into the feet, or hands and fingers, and collecting there. Two evils are thus avoided, namely, under-nutrition of the upper parts of the limbs, and a permanently swollen condition of the hands and feet. Valves were made, therefore, to ensure a really fair general distribution of the blood for the nutrition of the various parts. A discussion of these valves must be preceded by an expression of wonder at the way in which they have hitherto escaped the notice of Anatomists, both of our own and earlier generations; so much so that not only have they never mentioned, but no one even set eye on them till 1574, when to my great delight [summa cum laetitia] I saw them in the course of my dissection.

We have here a beautiful example of the distinctness of the rules of representation and the rules of inference in the case in which the scope of explanation is unambiguous. Fabricius finds it astonishing that the existence of the valves has escaped the attention of anatomists; we find it astonishing that he draws wrong inferences from their existence. However, the discovery of the *existence of venous valves* and their *relatively accurate representation* (Figure 7), even if the *wrong inferences* were drawn from it, was yet another step of providing a *correct ingredient for a successful explanation* of the blood circulation to be provided by his student, William Harvey.

Fabricius, himself a celebrity during his lifetime – a huge amphitheatre was especially constructed to his honour in Padua by the Venetian authorities – was only one among others such as Amatus Lusitanus, Eustachius, Sylvius who gave representations of the valves in the veins in the explanatory game and whose contributions will not be discussed here. However, I would like to draw the attention to Miguel Serveto, who in the fifth book of his religious treatise

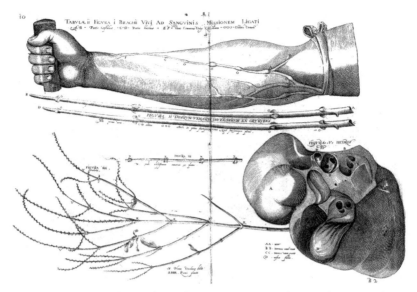

FIGURE 7 This is Plate II from Hieronymus Fabricius ab Aquapendente, *De Venarum Ostiolis*, 1603. Figura i is especially important here and this is the English translation of the "Explanatio" provided by Fabricius for it (p. 58):

Figure I shews a living arm, bound above with a ligature, as usually happens in blood-letting. In it there is seen part of the cephalic or humeral vein A.B., and part of the basilic or iecoraria C.D., then the vena Communis popularly called Median E.F., in which, as in other veins, valves O.O.O. are seen like so many knots. For this is the kind of picture the valves present in living arms when viewed from the outside.

Courtesy of Wellcome Library, London

Christianismi Restitutio (The Restoration of Christianity), described the small circulation quite accurately in an account of the passage of the blood from the heart to the lungs and then back again to the left ventricle of the heart. This work, however, though published already in 1553 and obviously based on studies having taken place earlier and, thus, earlier than Colombo's *Re Anatomica*, had no impact whatsoever in the explanatory process of the functioning of the heart and the blood circulation: having been found offensive to Catholics,

Lutherans and Calvinists, the author was condemned for spreading and preaching Nontrinitarianism and sentenced to death by burning at the stake in Geneva in 1553 with one of the apparently last copies of his book chained to his leg.[25] What is important here is that metaphysical assumptions that were irrelevant to the explanatory game prevented communication and inhibited the explanatory process by means of the physical extinction of both the one engaged in the explanatory activity and the product alike.

The explanation of the functioning of the heart and of the circulation of the blood that William Harvey provided in *De motu cordis* is concomitant with a complete change of nearly all kinds of rules of the explanatory game in comparison with those employed in Galenism. The most significant change concerning the *metaphysical rules* had to do with the abandonment of the postulation of different spirits. The standard *rules of representation* pertaining to the dissection of cadavers have been enriched by novel diagrammatic rules – most importantly the relatively accurate representation of the valves as offered by Fabricius. And the employment of mathematical calculations has become the decisive novel addition to the more standard rules of analogy that have long been employed in the field, constituting, thus, a change in the *rules of inference*. The *successful dissemination* of Harvey's explanation was due to the focused way of its presentation in the form of a slender volume. I will briefly address the employment of all these kinds of rules in Harvey's *Exercitatio anatomica de motu cordis et sanguinis in animalibus* starting with the metaphysical ones (1628, Engl. transl. by Robert Willis 1989, p. 116f.):

> Persons of limited information, when they are at a loss to assign a cause for anything, very commonly reply that it is done by the spirits; and so they bring the spirits into play upon all occasions; even as indifferent poets are always thrusting the gods upon the

[25] Of 1,000 copies of his book only three still exist, one in the National Library of Austria in Vienna, one in the Bibliothèque National in Paris, which was most likely used in his trial, and one in the library of the University of Edinburgh.

stage as a means of unravelling the plot, and bringing about the catastrophe. [...] They who advocate incorporeal spirits have no ground of experience to stand upon; their spirits indeed are synonymous with powers or faculties, such as a concoctive spirit, a chylopoetic spirit, a procreative spirit &c. – they admit as many spirits, in short, as there are faculties or organs. [...] [T]here is nothing more uncertain and questionable, then, than the doctrine of spirits [...]

Regarding the rules of representation the similarity of the only illustrations from the folding leaf included in Harvey's book with those contained in Fabricius's *De Venarum Ostiolis* is characteristic (Figure 8). These representations have helped decisively to provide the novel explanatory move regarding the circulation of the blood. According to the famous recollection of Boyle (1688, p. 157f.):

And I remember that when I asked our famous Harvey, in the only discourse I had with him, (which was but a while before he died), what were the things that induced him to think of a circulation of the Blood? He answered me, that when he took notice that the valves in the veins of so many several parts of the body, were so placed that they gave free passage to the blood towards the heart, but opposed the passage of the venal blood the contrary way: he was invited to imagine, that so provident a cause as nature had not plac'd so many valves without design: and no design seems more probable, than that, since the blood could not well, because of the interposing valves, be sent by the veins to the limbs, it should be sent through the arteries, and return through the veins, whose valves did not oppose its course that way.[26]

The novel rules of inference employed by Harvey included *the analogy of the heart as a pump* which decisively differed from the Galenic view of the blood which ebbs and flows in the arteries (Acierno 1994, p. 217). And very importantly also the use of (simple)

[26]   For a useful discussion of the recollections of Boyle see McMullen (1995, pp. 492ff.).

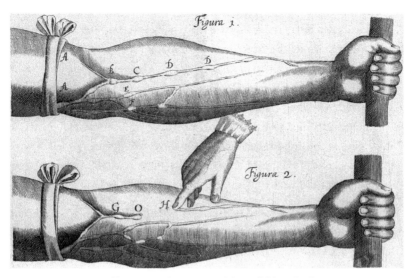

FIGURE 8 Illustration (Figures 1–2) from folding leaf in Harvey's
Exercitatio anatomica de motu cordis et sanguinis in animalibus from
the 1628 edition, Frankfurt: Sumptibus Guilielmi Fitzeri.
Courtesy of Wellcome Library, London

mathematical calculations which made the explanation of the *circulation* of the blood appear evident and emerge spontaneously (Harvey 1628 Engl. transl. by Robert Willis 1989, p. 48f.):

> Let us assume either arbitrarily or from experiment, the quantity of blood which the left ventricle of the heart will contain when distended to be, say two ounces, three ounces, one ounce and a half – in the dead body I have found it to hold upwards of two ounces. Let us assume further, how much less the heart will hold in the contracted than in the dilated state; and how much blood it will project into the aorta upon each contraction [...]; and let us suppose as approaching the truth that the fourth, or fifth or sixth, or even but the eighth part of its charge is thrown into the artery at each contraction; this would give either half an ounce or three drachms or one drachm of blood as propelled by the heart at each pulse into

the aorta; which quantity, by reason of the valves at the root of the vessel, can by no means return into the ventricle. Now in the course of half an hour the heart will have made more than one thousand beats, in some as many as two, three or even four thousand. Multiplying the number of drachms propelled by the number of pulses, we shall have either one thousand half ounces, or one thousand times three drachms, or a like proportional quantity of blood, according to the amount which we assume as propelled with each stroke of the heart, sent from this organ into the artery; a larger quantity in any case as is contained in the whole body!

The *integrated employment of all those novel rules* have led to a novel explanation of the blood circulation, summarized in what according to Khan, Daya and Gowda (2005, p. 521) is probably the most significant medical declaration ever published (Harvey 1628 Engl. transl. by Robert Willis 1989, p. 68):

And now I may be allowed to give in brief my view of the circulation of the blood, and to propose it for general adoption.

Since all things, both argument and occular demonstration, show that the blood passes through the lungs and heart by the action of the [auricles and] ventricles, and is sent for distribution to all parts of the body, where it makes its way into the veins and pores of the flesh, and then flows by the veins from the circumference on every side to the centre, from the lesser to the greater veins, and is by them finally discharged into the vena cava and right auricle of the heart, and this in such a quantity or in such a flux and reflux thither by the arteries, hither by the veins, as cannot possibly be supplied by the ingesta, and is much greater than can be required for mere purposes of nutrition; it is absolutely necessary to conclude that the blood in the animal body is impelled in a circle, and is in a state of ceaseless motion; that this is the sole and only end of the motion and contraction of the heart.

A crucial refinement of the rules of representation has been enabled by the invention and use of the microscope. It was due to

Malpighi's observation of the lungs of frogs with the aid of a microscope in 1661 that capillaries have been identified and, thus, the passage of the blood from the outermost branchings of the arteries to the outlying ramifications of the veins could be shown (see Figure 9). Blood circulation became an even more acceptable explanation of the functioning of the heart (Malpighi 1661/1929, p. 8):

> The power of the eye could not be extended further in the opened living animal, hence I had believed that this body of the blood breaks into the empty space, and is collected again by a gaping vessel and by the structure of the walls. The tortuous and diffused motion of the blood in diverse directions, and its union at a determinate place offered a handle to this. But the dried lung of the frog made my belief dubious. This lung had, by chance, preserved the redness of the blood in (what afterwards proved to be) the smallest vessels, where by means of a more perfect lens, no more there met the eye the points forming the skin called Sagrino, but vessels mingled annularly. And, so great is the divarication of these vessels as they go out, here from a vein, there from an artery, that order is no longer preserved, but a network appears made up of the prolongations of both vessels. This network occupies not only the whole floor, but extends also to the walls, and is attached to the outgoing vessel, as I could see with greater difficulty but more abundantly in the oblong lung of a tortoise, which is similarly membranous and transparent. Here it was clear to sense that the blood flows away through tubules, and is dispersed by the multiplex winding of the vessels. Nor is it a new practice of Nature to join together the extremities of vessels, since the same holds in the intestines and other parts; nay, what seems more wonderful, she joins the upper and the lower ends of veins to one another by visible anastomosis, as the most learned Fallopius has very well observed.

The acceptance of the use of novel rules of representation in the form of the microscopic studies was anything other than straightforward or unproblematic. In fact, the contrary was the case. Although Malpighi

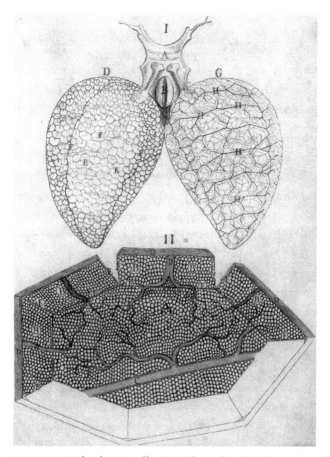

FIGURE 9 Malpighi Marcello, *De Pulmonibus* (1661/1929, p. 11) showing the lungs of a frog with a cross-sectional microscopic view.
Courtesy of Wellcome Library, London

was a very modest and gentle man, the immense scope and impact of his microscopic studies provoked such envy and criticism, that in 1684 his villa was burnt by adversaries; his papers, notes and manuscripts destroyed; and his laboratory equipment ruined.

Closing, it is important to stress that an explanatory game had been unfolding in the Arab world in parallel to the one in the West already in the thirteenth century. The Arabian physician Ibn an-Nafis (1213–1288) had rejected the Galean view of the existence of interventricular pores and maintained that only via the lungs could the blood reach from the right ventricle to the left ventricle of the heart.[27] In his *Commentary on Anatomy in Avicenna's Canon*, in the English translation made by Meyerhof (1935, p. 116) of the section of the book by Ibn al-Nafis identified as fol. 46 r, one can read:

> [. . .] This cavity is the right cavity of the heart. The blood, after it has been refined in this cavity, must be transmitted to the left cavity where the [vital] spirit is generated. *But there is no passage between these two cavities*; for the substance of the heart is solid in this region and has neither a visible passage, as was thought by some persons, nor an invisible one which could have permitted the transmission of blood, as was alleged by GALEN. The pores of the heart there are closed and its substance is thick. Therefore, *the blood*, after having been refined, *must rise in the arterious vein to the lung in order to expand in its volume and to be mixed with air* so that its finest part may be clarified *and may reach the venous artery in which it is transmitted to the left cavity of the heart.* This, after having been mixed with the air and having attained the aptitude to generate the [vital] spirit. That part [of the blood] which is less refined, is used by the lung for its nutrition.

In the immediately following short passage, Ibn al-Nafis seems to be the first one to speak of the existence of capillaries reported by Malpighi 400 years later with the aid of his microscopic studies on the lungs of frogs:

> For this reason the arterious vein has solid substance with two layers, in order to make more refined that (the blood) which transsudes from it. The venous artery, on the other hand, has thin substance in order to facilitate the reception of the transsuded

---

[27]   On Ibn an-Nafis see the monograph by Fancy (2013).

[blood] from the vein in question. And for the same reason there exists perceptible passages (or pores, *manāfidh*) between the two [blood vessels].

However, there had been apparently no dissemination of these descriptions in the Western world until 1922 when the *Commentary* was discovered by an Egyptian medical student in the Prussian State Library (Meyerhof 1935, p. 102). This is an excellent exemplification of what we will discuss in the next chapter under the heading of the plurality of explanatory games in a horizontal dimension: at every moment of time there is a continuum of explanatory knowledge available; this knowledge is produced while diverse communities proceed in explanatory activities. Thus, many explanatory games take place concurrently.

The discussion of the case studies lends support to the claim that during the long historical duration (especially the historical period during which a research area is developed) the same explanatory game is being played, which is appropriately transformed in order to include new data and possibly new outlooks of the same subject. Hence, the explanatory game, remaining itself the same since it refers to the same phenomenon, changes diachronically adapting itself to the historically created epistemological framework which is characterized both by the nature of the represented phenomenon and by the socially and historically dependent viewpoint of the cognizer. Explanatory knowledge, neither absolute nor relativistically anarchic, has a historical and social character being firmly grounded, however, in the indisputable reality of the phenomena.

7 The Plurality of Explanatory Games

At every moment of time there is a plurality of explanatory games in a society that take place in parallel. In other words, it is not only within science that different explanatory games are played but also within society at large. Lay people use rules of representation, rules of inference and rules of scope in order to provide explanations about phenomena that interest them. Children permanently ask why-questions to their parents and receive responses that are aimed to satisfy their curiosity. The vast majority of explanations offered in a society are common-sense explanations by lay people, following quite rudimentary rules in their explanatory activities. However, these explanations very often, though by no means always, tend to change, and under appropriate circumstances they become more and more convincing. In other words, the explanatory labour is not only distributed among the different participants in the diverse games at a moment of time but they also evolve over time. This evolution takes place as the sets of rules that guide the activities of the explainers change over time. Thus, the broad picture is one of both a horizontal division of explanatory labour at a moment of time and a vertical division of explanatory labour over time. The horizontal division can be best conceptualized as a plurality of explanatory games existing in parallel at a time, and the vertical division as the rule-change of these games over time. I will elaborate further on this in what follows.

7.1 THE HORIZONTAL DIMENSION

At every moment of time there is a continuum of explanatory knowledge available in a society. This knowledge is produced while diverse communities (broadly defined to include any group of people) proceed in explanatory activities. Thus, many explanatory games take place

97

concurrently. This is trivial, if one considers that one set of the constitutive rules of any explanatory game determines what counts as an explanandum. Human beings wish to explain a huge number of phenomena in their environment, so it is only natural to postulate that there, accordingly, must be a huge number of games. What makes the claim of plurality more interesting, however, is that the constitutive rules are also comprised by rules determining what must be taken as given and by rules determining the metaphysical presuppositions of the game. So, we can have explanatory games which have the same explananda, but differ in the rules determining what must be taken as given and/or what the metaphysical presuppositions are. Here the plurality becomes more interesting, because we can identify different explanatory games that deal with the same explananda, but which differ from one another on the basis of what they assume as given and of their metaphysical assumptions.

Three important types here are mythical, religious and scientific explanatory games. There are cases in which these three types of explanatory games simply deal with distinct and different explananda and, thus, there is no overlap among them (consider the question of why angels need to have wings in order to fly). In many cases, however, these three types of games deal with the same explananda, but differ radically in what they take as given and in their metaphysical presuppositions. Consider the case of the emergence of the universe, cosmogony.

A prominent explanation that was produced in the framework of the Greek mythical explanatory game is best summarized in the following quotation from Hesiod's *Theogony* (114–138):[1]

> Tell me these things, Olympian Muses, tell
> From the beginning, which first came to be?
> Chaos was first of all, but next appeared
> Broad-bosomed Earth, sure standing-place for all
> The gods who live on snowy Olympus' peak

---

[1]   This is the translation in the Penguin Classics edition (1976). For a useful discussion of cosmogonic myths see Burkert (1999). An excellent discussion is also provided by Topitsch (1958, pp. 5–95).

And misty Tartarus, in a recess
Of broad-pathed earth, and Love, most beautiful
Of all the deathless gods. He makes men weak,
He overpowers the clever mind, and tames
The spirit in the breasts of men and gods.
From Chaos came black Night and Erebos.
And Night in turn gave birth to Day and Space
Whom she conceived in love to Erebos.
And Earth bore starry Heaven, first, to be
An equal to herself, to cover her
All over, and to be a resting-place,
Always secure, for all the blessed gods.
Then she brought forth long hills, the lovely homes
Of goddesses, the Nymphs who live among
The mountain clefts. Then, without pleasant love,
She bore the barren sea with its swollen waves,
Pontus. And then she lay with Heaven, and bore
Deep-whirling Oceanus and Koios; then
Kreius, Iapetos, Hyperion,
Theia, Rhea, Themis, Mnemosyne,
Lovely Tethys, and Phoebe, golden-crowned.
Last, after these, most terrible of sons,
The crooked-scheming Kronos came to birth
Who was his vigorous father's enemy.

A prominent explanation of cosmogony that was produced in the framework of the Christian explanatory game is mainly contained in the Book of Genesis, the first chapter of which I would like to quote at length:

> In the beginning when God created the heavens and the earth, the earth was a formless void and darkness covered the face of the deep, while a wind from God swept over the face of the waters. Then God said, "Let there be light"; and there was light. And God saw that the light was good; and God separated the light from the darkness. God

called the light Day, and the darkness he called Night. And there was evening and there was morning, the first day. And God said, "Let there be a dome in the midst of the waters, and let it separate the waters from the waters." So God made the dome and separated the waters that were under the dome from the waters that were above the dome. And it was so. God called the dome Sky. And there was evening and there was morning, the second day. And God said, "Let the waters under the sky be gathered together into one place, and let the dry land appear." And it was so. God called the dry land Earth, and the waters that were gathered together he called Seas. And God saw that it was good. Then God said, "Let the earth put forth vegetation: plants yielding seed, and fruit trees of every kind on earth that bear fruit with the seed in it." And it was so. The earth brought forth vegetation: plants yielding seed of every kind, and trees of every kind bearing fruit with the seed in it. And God saw that it was good. And there was evening and there was morning, the third day. And God said, "Let there be lights in the dome of the sky to separate the day from the night; and let them be for signs and for seasons and for days and years, and let them be lights in the dome of the sky to give light upon the earth." And it was so. God made the two great lights – the greater light to rule the day and the lesser light to rule the night – and the stars. God set them in the dome of the sky to give light upon the earth, to rule over the day and over the night, and to separate the light from the darkness. And God saw that it was good. And there was evening and there was morning, the fourth day.

Contemporary cosmology, the branch of physics that deals with the question of the emergence of the universe, proposes a big bang explanation. According to the general theory of relativity, gravity is not a force but the curvature of space time. The range of gravity is infinite since it is a property of space time itself and the evolution of the universe is ultimately dictated by gravity. A set of differential equations relates the dynamic quantities in the universe and turns them back in time. Simulations on computers relate initial conditions of the universe to different outcomes aiming at representations of the kind of universe we live in. Earlier astronomers, most prominently Edwin Hubble, observed that neighbouring galaxies of the Milky Way galaxy are receding away – the more distant they are, the faster were they found to be moving away. In other words, recessional velocity of a galaxy increases with its distance from the earth. Reversing this expansion scenario back in time, the inference is drawn that if galaxies are moving rapidly apart now, they must have been denser – with more matter and energy per unit volume – in the past. Going back in time, the whole universe should converge to a point of infinite density and extremely high temperatures, and this should be the starting point of big bang. The universe itself should have been an infinitely dense point which expanded to its present size. The age of the universe according to this Big Bang explanation is estimated to be 13.73 billion years.

Given that Hesiod was active between 750 and 650 BCE and that his mythical explanation has been reproduced for centuries afterwards and that Genesis was composed in the late seventh or the sixth century BCE (according to the prevailing consensus of biblical scholars), the first two explanatory games have been unfolding in parallel for a long period of time. Today, nobody believes in the mythical explanation of Hesiod. Today, it is the religious explanatory game of the Genesis, along with the scientific explanatory game of cosmology, that are unfolding in parallel. (And there are other religious explanatory games that deal with the same explanandum, which I leave unmentioned here.)

The biblical and the scientific explanatory games are structured around the same explanandum, but differ radically in the rules determining what must be taken as given and the rules determining their metaphysical presuppositions. In their everyday work, though the players of the respective games engage in their explanatory activities following the constitutive rules of their own game, they occasionally engage critically with the constitutive rules of the other games. This engagement is usually manifest in the rejection of what is taken as given by the players of the other game and a rejection of the metaphysical presuppositions that they employ. The ongoing debate on creationism is such an example.

7.2   THE VERTICAL DIMENSION

Explanatory games evolve over time. An important factor of their evolution is that even the most fundamental of the constitutive rules, the ones that determine what counts as an explanandum, evolve over time. The legitimate why-questions to ask in a specific domain change over time, something that should not be astonishing.[2]

The change in the constitutive rules is certainly important, but the prevalent change of all other kinds of rules is of equal importance, making the playing of an explanatory game an inherently dynamic enterprise. The means of representing phenomena evolve over time – in contemporary science, artefacts like graphs, computer monitor displays, etc. are constantly developed further as are abstract means of representation like mathematical tools. The same is the case with the rules of inference: this is what the traditional philosophy of science has focused on, debating the nature of these rules, that is, whether they are lawful, and if so, what their lawfulness consists in, whether they function as mere inference tickets from data to data or whether one

---

[2]   Thomas Kuhn (1962/1970, pp. 106ff.) has drawn attention to that and van Fraassen (1980, p. 111f.) has raised the issue again in the theory of explanation: "Examples [Kuhn] gives of explanation requests which were considered legitimate in some periods and rejected in others cover a wide range of topics. They include the qualities of compounds in chemical theory (explained before Lavoisier's reform, and not considered something to be explained in the nineteenth century, but now again the subject of chemical explanation). Clerk Maxwell accepted as legitimate the request to explain electromagnetic phenomena within mechanics. As his theory became more successful and more widely accepted, scientists ceased to see the lack of this as a shortcoming. The same had happened with Newton's theory of gravitation which did not (in the opinion of Newton or his contemporaries) contain an explanation of gravitational phenomena, but only a description. In both cases there came a stage at which such problems were classed as intrinsically illegitimate, and regarded exactly as the request for an explanation of why a body retains its velocity in the absence of impressed forces. While all of this may be interpreted in various ways (such as through Kuhn's theory of paradigms) the important fact for the theory of explanation is that not everything in a theory's domain is a legitimate topic for why-questions; and that what is, is not determinable *a priori*."

should interpret them realistically, whether the different ontological structures require different kinds of law-like generalizations, etc. Finally, the rules of scope also change over time: new instructions about where and how to apply the explanatory rules to new phenomena are issued, debated and followed by the participating scientists.

The explanations offered by the participants of an explanatory game evolve over time as they follow the different kinds of rules available to them. Explanatory knowledge is just a part of general human knowledge that is a product of genetic and cultural evolution (Sterenly 2014). The process of cultural evolution concerns the change and transmission of knowledge in time at a societal level, and it can be regarded as a process of collective learning (Mantzavinos 2001, p. 73). Donald (1991), in his *Origins of the Modern Mind*, highlights the function of what he calls "External Symbolic Storage" for the trans-mission and accumulation of knowledge across generations. It was the simple habit of recording ideas, that is, "of externalizing the process of oral commentary of events" (p. 342), that constituted the critical innovation that has massively supported the evolution of theoretic culture. This occurred in Greece beginning around 700 BCE. What was truly innovative was that "for the first time in history complex ideas were placed in the public arena, in an external medium, where they could undergo refinement over the longer term, that is, well beyond the life-span of single individuals" (p. 344). These External Symbolic Storage Networks have decisively changed the character of knowledge evolution, since they provide the possibility of a constant interaction between the accumulated corpus of theoretical knowledge and the theoretical problems of the individuals in a society.

If one adopts an evolutionary perspective,[3] then the thesis of a continuum between everyday, commonsensical knowledge and scien-tific knowledge, both products of the more general cultural evolution-ary process, seems very plausible. The thesis of a continuum does not deny the difference in the quality of the scientific knowledge vis-à-vis

---

[3]   For an evolutionary account of science, see Hull (1988).

other kinds of knowledge. But instead of focusing on clear-cut demarcation criteria of a *syntactic nature* (something that was fashionable
for a long time in philosophy of science), it suggests that it is the
specific way that knowledge is imposed for criticism that one should
rather study. An institutional arrangement that enables criticism is
the first prerequisite, of course. With this I mean that the institutional
structure, both in its formal and in its informal elements, must allow
or even encourage the criticism of knowledge structures and practices.
A polity that guarantees the freedom of expression and allows for
resources to be devoted to the building of organizational structures
such as universities, where everything is allowed to be questioned,
provides the formal institutional structure that makes advanced criticism possible. A widespread critical attitude towards given beliefs and
practices provides the informal institutional structure that enables
criticism. The latter is more important (and has in fact come first
historically).[4] The specific mix of formal and informal institutions
prevailing in a society regulates the behaviour of the individuals and
provides the solution to what is (very inaccurately) called the problem
of "power".[5]

In theocratic systems there is strong resistance to the criticism of
religious explanations. Communist political systems have similarly
regularly disallowed the development of explanatory games for the

---

[4]   See Popper (1963/1989, p. 50): "For the critical attitude is not so much opposed to the
dogmatic attitude as super-imposed upon it: criticism must be directed against existing
and influential beliefs in need of critical revision – in other words, dogmatic beliefs. A
critical attitude needs for its raw material, as it were, theories of beliefs which are held
more or less dogmatically. Thus science must begin with myths, and with the criticism
of myths; neither with the collection of observations, nor with the invention of
experiments, but with the critical discussion of myths, and of magical techniques
and practices. The scientific tradition is distinguished from the pre-scientific tradition
in having two layers. Like the latter, it passes on its theories; but it also passes on a
critical attitude towards them. The theories are passed on, not as dogmas, but rather
with the challenge to discuss them and improve upon them. This tradition is Hellenic:
it may be traced back to Thales, founder of the first school (I do not mean 'of the first
philosophical school', but simply 'of the first school') which was not mainly concerned
with the preservation of a dogma."

[5]   I have developed my own theory of institutions which gives what I think is a more
accurate account of how "power" – indeed a very vague term – is regulated in a society.
See Mantzavinos (2001).

analysis of social phenomena other than the ones that the communist party has found acceptable. The range and amount of criticism thus decisively depends on the prevailing institutional framework – the freedom of thought and expression that enables scientific explanatory activities in modern universities and is guaranteed by liberal states around the world is only a contingent historical case.

Thus, explanatory games unfold through time in the context of specific institutional constraints, and in the broader context of an ongoing evolution of all human knowledge. The rules of representation, the rules of inference and the rules of scope followed by the participants of any specific game change over time. Since the ingredients of explanations come from the use of these different kinds of rules, their change over time is concomitant with a vertical division of explanatory labour. The broad picture of the division of explanatory labour consists, thus, both of a horizontal dimension comprising all explanatory games at a moment of time and of a vertical dimension encapsulating all changes of explanatory activities taking place within the diverse games over time. How exactly these rules change over time is, of course, the crucial issue with which we will deal in the next chapter.

What should have emerged as an intermediate result of the analysis so far, however, is the following: it is very fruitful and productive to conceive of explanation in terms of games since both the evolutionary character of explanatory activity unfolding in historical time and its social nature are highlighted appropriately, while at the same time the phenomena themselves remain stable. To recapitulate and precisify: explanatory games evolve over time. This claim implies that the games survive through time while evolving, which goes hand in hand with the more groundbreaking claim regarding the stability of the phenomena to be explained. Even the constitutive rules change in their entirety while evolving. That is, change refers not only to the rules which determine what must be taken as given and to the rules determining the metaphysical presuppositions but also to the *rules determining what counts as an explanandum*. But if the explanandum itself is changing, what does remain stable? The phenomena themselves to be explained! This which changes and which makes up the changing explanandum are not the phenomena themselves, but their representation, that is, our cognitive access into them. This which changes and which is the target of explanation are not the phenomena themselves but our image of them which possesses a historic and social character.

8 Explanatory Activity as Problem-Solving Activity

The rules of the explanatory game emerge and change in a process of a spontaneous interaction between the participants of the game. The nature and the specific characteristics of this process will be analysed in detail in the next chapter. In this chapter the task will be to conceptualize explanatory activity as problem-solving activity and to provide a descriptive account of the explainer as a problem-solver. I will focus first exclusively on the individual, before tackling in a second step the social interaction between individuals engaging in explanatory activities.

8.1 THE PROBLEM-SOLVING FRAMEWORK

One fruitful way to view human activity is as problem-solving activity – human beings are constantly confronted with problems and they mobilize their cognitive and emotional resources in order to solve them. Cognitive science has developed a series of approaches to capture problem-solving activities in different domains and contexts. Newell and Simon, in their classic *Human Problem Solving*, have outlined the breadth of the approach, which encompasses all aspects of human activity (1972, p. 72):

> A person is confronted with a problem when he wants something and does not know immediately what series of actions he can perform to get it. The desired object may be very tangible (an apple to eat) or abstract (an elegant proof for a theorem). It may be specific (that particular apple over there) or quite general (something to appease hunger). It may be a physical object (an apple) or a set of symbols (the proof of a theorem). The actions involved in obtaining desired objects include physical actions (walking, reaching, writing), perceptual activities (looking, listening), and purely

mental activities (judging the similarity of two symbols, remembering a scene, and so on).

Problem solving can be conceptualized as a process of searching through a state space: Holland *et al.* (1986, p. 10) define a problem as an *initial state*, by one or more *goal states* to be reached by a set of *operators* that can transform one state into another and by *constraints* that a solution must meet. The problem-solving procedure can be conceptualized, thus, as a series of steps in which operators are applied so that the goal states can be reached – the set of operators are information processes, each producing new states of knowledge (Newell and Simon 1972, p. 810). Apart from this specific kind of conceptualization of a problem-solving activity, which is indeed very valuable since it has led to a series of workable models of inference and learning, it is the core of the approach that is important here. Most, if not all, phenomena of the mind can be viewed in the context of problem solving: human beings perceive the world, learn from their environment and act according to their problem situation. There is no perception per se, but always perception in relation to a problem; there is no learning per se, but always learning about ways to solve problems.

This view is not only prevalent in cognitive science (Mayer 1992, Davidson and Sternberg 2003) but is also largely shared by different philosophical traditions. Apart from the more obvious tradition of pragmatism, problem solving is endorsed by evolutionary epistemology in the footpaths of Lorenz,[1] Campbell[2] and Popper[3] and is at the core of Laudan's classic work *Progress and its Problems* (1977).[4] The problem-solving framework is fruitful because it includes

---

[1]   See especially Lorenz (1941, 1943, 1965, 1973).
[2]   See especially Campbell (1965, 1974/1987).
[3]   See especially Popper (1972/1992, 1987). Though Popper's later work seems to belong to this tradition, I would like to stress the fact that he himself did not accept this characterization in a remark that he made in Popper (1987, p. 32). For a thorough account of the approach of evolutionary epistemology in the German-speaking world see Vollmer (1975/2002).
[4]   See the opening phrases of the first chapter of this book of Laudan (1977, p. 11): "Science is essentially a problem-solving activity. This anodine bromide, more a cliché than a philosophy of science, has been espoused by generations of science textbook writers

an analysis of the context – the narrowly epistemological but also the broader social and institutional one. Besides, it is concomitant with a dynamic account of knowledge, it allows problem solutions to be evaluated in reference to degrees of adequacy, and most importantly it allows for different criteria regarding what should count as an adequate solution to a problem. Depending on the focus of interest, the problem-solving framework can be applied to a variety of issues – also in epistemology, which primarily concerns us here.

Since Ryle's (1949) criticism of the common intellectualist legend that all of our acting is guided by some precedent thought operation and his introduction of the dichotomy between "knowing that" and "knowing how", philosophers and cognitive scientists alike have made considerable progress in theorizing on the more specific ramifications of this distinction. "Knowledge that" emerges as the outcome of solving "theoretical problems" and "knowledge how" as the outcome of solving "practical problems"; and a useful criterion for distinction is the level of *communicability* of knowledge (Mantzavinos 2001, pp. 30ff.). "Knowledge that" can be communicated by means of language using a series of symbols, whereas skills and arts cannot be communicated by symbolic languages. The most common way that "knowledge how" can be shared with other persons is through learning by example, that is, learning by imitation. The process of learning how to perform certain tasks is thus not always expressible in linguistic terms and is therefore mainly communicable through direct imitation.[5]

---

and self-professed specialists on '*the* scientific method'. But for all the lip service which has been paid to the view that science is fundamentally the solving of problems, scant attention has been paid, either by philosophers of science or historians of science, to the ramifications of such an approach for understanding science."

[5] See Ryle (1949, p. 41): "We learn how by practice, schooled indeed by criticism and example, but often quite unaided by any lessons in the theory."

See Polanyi (1958, p. 62): "The unspecifiability of the process by which we thus feel our way forward accounts for the possession by humanity of an immense mental domain, not only of theoretical knowledge but of manners, laws and of many different acts which man knows how to use, comply with, enjoy or live by, without specifiably knowing their contents."

The crucial argument in favour of establishing the distinction between "knowing that" and "knowing how" seems to be as valid as it was when Ryle first formulated it. It is as follows: According to a familiar myth, before any action, we consider how we should perform our task. An intelligent action is, then, one in which the agent has been involved in thinking through what she is doing while undertaking the action and one in which contemplation of the appropriate actions has decisively promoted her success. As Ryle put it: "To do something thinking what one is doing is, according to this legend, always to do two things; namely to consider certain appropriate propositions, or prescriptions, and to put into practice what these propositions or prescriptions enjoin. *It is to do a bit of theory and then to do a bit of practice*" (1949, p. 30). However, the consideration of the propositions upon which the action is supposed to be based is itself an operation that can succeed only if this consideration has been more or less intelligent. And if any operation to be executed intelligently or successfully requires a prior theoretical operation, then we have a logical circle that nobody can evade (Ryle 1949, p. 31). It seems plausible, thus, that "knowing how" does not require "knowledge that".[6]

The distinction between declarative knowledge and procedural knowledge is a largely accepted one in cognitive science, and it is generally acknowledged that learning facts is very different than

---

Merlin Donald (1991, ch. 6), argues for a distinct evolutionary episode in the transition from ape to human that he calls "mimetic culture" and which is still existing in modern culture.

For an excellent review of the modern discussions on the issue of tacit and implicit knowledge from the perspective of modern epistemology see Davies (2015).

[6] See another statement of the position by Ryle (1949, p. 31): "Efficient practice precedes the theory of it; methodologies presuppose the application of the methods, of the critical investigation of which they are the products. It was because Aristotle found himself and others reasoning now intelligently and now stupidly and it was because Izaak Walton found himself and others angling sometimes effectively and sometimes ineffectively that both were able to give to their pupils the maxims and prescriptions of their arts. It is therefore possible for people intelligently to perform some sorts of operations when they are not yet able to consider any propositions enjoining how they should be performed. Some intelligent performances are not controlled by any interior acknowledgments of the principles applied in them."

acquiring skills.[7] Finally, neurological studies of patients suffering from amnesia have shown that the difference between "knowing how" and "knowing what" is honoured by the nervous system. In a classic study, for example, Cohen and Squire (1980) report on patients who were capable of acquiring a "mirror-reading skill" although they neither had a memory of the words they read nor even of being confronted with the task. Their amnesia in relation to "knowing that" (in this case, of the specific words and the fact that they dealt with them in a laboratory experiment) did not hinder the learning or the exercising of "knowing how" (in this case, the reading of words that were presented in mirror images).

The distinction between declarative knowledge and procedural knowledge will prove important for the account of explanatory activity; but before proceeding to an analysis of the explainer as a problem-solver, some further elaboration is required.

8.2   OLD AND NEW PROBLEMS

The problem-solving framework that I have outlined very generally can be specified more closely with the help of rule-based theories of problem solving of contemporary cognitive science. According to the

---

[7] For an overview see Anderson (2010, ch. 9). Stanley and Williamson (2001) have contested the view that "knowing that" and "knowing how" are two distinct *kinds* of states of knowledge and they defend the view that all "knowing how" is "knowing that". This heretic position has been heavily criticized (see for example Schiffer 2002; Koethe 2002; Rosefelt 2004). In their original contribution Stanley and Williamson did not consider any kind of experimental evidence produced by cognitive science, since theirs has been an exercise in epistemology of the analytical genre. Stanley in chapter 7 of his book on *Know How* discusses the findings of cognitive science, but reinterprets the classic distinction between declarative and procedural knowledge in such a way as to fit his own view: "The debate about declarative knowledge and procedural knowledge does not concern the existence of states of knowledge that lack truth-evaluable consent. It is rather a dispute about how to *implement knowledge*, i.e. how best to derive propositional knowledge states, procedurally or declaratively" (Stanley 2011, p. 152). Such an interpretation of the masses of empirical evidence made available over the last decades in cognitive science is hardly convincing, of course. For an emphatic criticism of the approach of Stanley and Williamson and more specifically of their choice to thoroughly neglect dozens of very diverse empirical studies see Devitt (2011). Wallis (2008) and Adams (2009) also discuss a series of empirical studies from neuroscience and experimental psychology and convincingly show that Stanley and Williamson are in error.

rule-based approach, the mind forms internal representations in the form of "problem space" (Newell and Simon 1972, p. 59) and perception and learning take place pragmatically. The classification of events and their interpretation in the light of a current problem occurs according to classifier rules of the general "IF...THEN" type. A main characteristic of the human mind is its infinite potential to create new rules, that is, to group items into classes. A number of rules and rule clusters organized in default hierarchies (Holland *et al.* 1986, p. 18) give rise to mental models. One plausible way to account for mental representations is, thus, by way of mental models which are coherent but transitory sets of rules that enable the mind to form predictions on what will occur in the environment, based on the available knowledge.[8] It is, on the one hand, essential to stress the *provisional, temporary* character of the mental models which differentiates them from other forms of mental representations proposed, like traditional rigid concepts, schemas (Bartlett 1932; Piaget 1936) and scripts (Schank and Abelson 1977), since they are *flexible knowledge structures created anew every time from the ready-made material of cognitive rules.* On the other hand, since mental models are the final prediction that the organism makes about the environment before getting feedback from it, it is their *fundamentally hypothetical* character that must be emphasized.

The building blocks of the mental models, that is, the IF...THEN rules, are in active competition, since many potential rules can guide thought and action in the environment. Rules whose conditions, that is, their IF-clauses, are satisfied by current messages compete to represent the current state of the environment. There are four criteria that are used to select the rules which will prevail in competition with the rest: "Thus, competition will favour those rules that (a) provide a description of the current situation (match), (b) have

---

[8]   Nersessian (2008a, p. 93) describes a mental model as a "structural, behavioral, or functional analog representation of a real-world or imaginary situation, event or process. It is analog in that it preserves constraints inherent in what is represented". See also Johnson-Laird (1983, 2006) and the list of the hundreds of articles on mental models at the website: www.mentalmodelsblog.wordpress.com.

a history of past usefulness to the system (strength), (c) produce the greatest degree of completeness of description (specificity) and (d) have the greatest compatibility with other current active information (support)" (Holland *et al.* 1986, p. 49).

This rule-based account does not only offer a plausible view of mental representations, but also a view of human learning so that there is a certain continuity in accounting for mental phenomena: learning takes place when the expectations formed are not successful in predicting the environment and must, thus, be modified. The environmental input triggers the creation of new mental models that may provide better expectations and guides for action: *learning is the complex process of rule modification according to the feedback received from the environment.* Learning is a dynamic process of trial and error: whenever an error occurs, a rule modification takes place, a new expectation is formed and, thus, a new trial is developed in order to solve the problem at hand. What is theorized as conceptual change in traditional philosophy of science is accounted for in the rule-based framework as a change of mental models. We will return to this later.

But, of course, there are not only errors: when a solution to a specific problem is obtained, the rules that led to the solution will be strengthened. Success in finding a solution is due to the success of the respective rules; and an apportionment of credit to the different rules that helped the organism to solve his problem is bound to happen (Holland *et al.* 1986, p. 70). If the same rules often led to acceptable solutions in the past, then they will repeatedly be employed by the organism in the future.

A series of successful solutions to the same problem creates what we call a *routine*. The essential characteristic of a routine is that it is employed to solve a problem without any prior reflection, that is, without conscious effort and the mobilization of cognitive resources. The use of routines is thus a function that does not require consciousness. Routines employed unconsciously give rise to rule-following behaviour; and the crucial fact is that since this behaviour was often followed in the past, it has become standardized.[9] Hence, once rules have been employed successfully many times to solve a problem, they are successively strengthened and stored in the memory; and after a time they take the form of unconscious routines.[10] This allows the limited cognitive faculties of the mind to be used thriftily, that is, consciousness needs to be concerned with problems that are difficult to solve or with new problems.[11]

A distinction between old and new problems is very useful in this context. Every time an individual is confronted with a problem situation, the human mind actively interprets it and classifies it. This presupposes, of course, that respective available classes already exist under which the current messages of the environment can be classified. When these classes do in fact exist, we can call the relevant problem situation an "old problem". These are the ones that can be classified under an already existing class that prescribes the appropriate solution (in the form of an IF...THEN rule). If the current problem is identified as a familiar one – in the sense that it can be classified in an existing class – then the appropriate solution will be applied automatically. Old problems are the ones that have been encountered

---

[9]   As Schrödinger correctly pointed out (1958, p. 7): "The fact is only this, that new situations and new responses they prompt are kept in light of consciousness and well practised ones are no longer so [kept]."

[10]   See Bargh and Chartrand (1999, p. 469): "But what we find most intriguing, in considering how mental processes recede from consciousness over time with repeated use, is that the process of automation itself is automatic. The necessary and sufficient ingredients for automation are frequency and consistency of use of the same set of component mental processes under the same circumstances – regardless of whether the frequency and consistency occur because of a desire to attain a skill, or whether they occur just because we have tended in the past to make the same choices or to do the same thing or to react emotionally or evaluatively in the same way each time. These processes also become automated, but because we did not start out intending to make them that way, we are not aware that they have been and so, when that process operates automatically in that situation, we aren't aware of it."

[11]   This solid empirical finding has been presaged by Whitehead already in 1911 (p. 45f.): "It is a profoundly erroneous truism, repeated by all copy-books and by eminent people making speeches, that we should cultivate the habit of thinking of what we are doing. The precise opposite is the case. Civilization advances by extending the number of operations which we can perform without thinking about them. Operations of thought are like cavalry charges in a battle – they are strictly limited in number, they require fresh horses, and must only be made at decisive moments."

frequently in the past, and their solution has been shown to be successful in the respective environment many times in the past. The solutions to these old problems consist in unconscious routines.

But there is no guarantee that a solution that has worked well in the past will prove successful in solving the problem at hand – this is the fundamentally hypothetical kernel of the approach. Whenever a solution to an old problem no longer works (possibly because of a change in the environment, i.e., the context) or whenever a problem situation, when compared to past problem situations stored in memory, cannot be classified under any familiar class, we speak of a "new problem". Naturally, all the old problems of an individual were new at some point in its history, so it is important to state what happens when an individual conceives of a situation as a new problem.

There are two sequential kinds of response to new problems. The first response is quasi-automatic and it involves the employment of so-called inferential strategies. These are processes of inference widely discussed in cognitive science mainly with respect to theoretical problems; they refer to processes of reasoning by which we draw a conclusion on the basis of available evidence or prior knowledge. Cases of inference such as the description and characterization of events, the detection of covariation among events, the use of causal inference, prediction and theory testing (Nisbett and Ross 1980, p. 10) are normally solved by lay persons with the aid of judgemental strategies or heuristics. Heuristics are to be understood as general strategies that provide quick solutions with little effort (Gigerenzer, Todd and the ABC Group 1999; Gigerenzer *et al.* 2011); they are meta-rules or higher-order rules that direct the employment of more specific rules in order to solve current problems. They are to be contrasted to "algorithms", that is, rigid methodical procedures that guarantee success by solving problems through their lengthy, patient application.[12]

---

[12]   See Wimsatt (2007, p. 10): "Heuristic principles are most fundamentally neither axioms nor algorithms, though they are often treated as such. As a group, they have distinct and interesting properties. Most importantly, they are re-tuned, re-modulated, re-contextualized, and often newly re-connected piecemeal rearrangements of existing

A common characteristic of heuristics is that they often will not successfully solve the problem for which they are used. This may be due to the nature of the problem or the inappropriate use of meta-rules. In either case, heuristics may lead to errors and the theoretical problems may not be adequately solved.

Despite their neglect in the literature, heuristics can also be used to solve practical problems. So, we can speak of inferential strategies with respect to mobilizing practical or "effector" rules (Holland *et al.* 1986, p. 42f.), perhaps the most powerful being *analogy* (Thagard 1996, ch. 5). If one has found a satisfactory solution to a problem in one domain, then an analogical transfer may lead to an equally good solution in another domain. If one knows how to ride a bicycle and she sits for the first time on a motorcycle, she will try to balance the same way as when she sat on a bicycle.

Hence, the first step that one takes to solve a new problem, either theoretical or practical, is to employ inferential strategies of a heuristic nature, according to which the employment of more specific rules is arranged. These inferential strategies are to be conceived as being the first intuitive response to any problem that is interpreted as a new one. There are, then, two obvious cases to be distinguished. Either the inference is successful, and accordingly the inferential meta-rules are strengthened and applied anew in the future, or it leads to a failure. In the second case, a mobilization of cognitive resources in the form of attention takes place and the second stage of the problem-solving process ensues.

This is a deliberation process, that is, a mental probing of alternatives and a choice is made of the one that is expected to best solve the new problem. In contrast to problems that are solved by routines, problem solving through a deliberative process is a conscious process.[13]

adaptations or exaptations, and they encourage us to do likewise with whatever we construct." See also Hey (2014).

[13] See Popper and Eccles (1977/1983, p. 126): "But the role of consciousness is perhaps clearest where an aim or purpose (perhaps even an unconscious or instinctive aim or purpose) can be achieved by alternative means, and when two or more means are tried out, after deliberation. It is the case of making a new decision."

Many problems that are interpreted as new ones can be solved by adopting ready-made solutions from the environment. Those ready-made solutions supply one of the alternatives that can be chosen when confronted with a new problem and the mind *creates* all the other alternatives ex nihilo. In an imaginative process, the mind devises new alternatives – creativity is the property of the mind that is exemplified when working out new alternatives in attempting to solve a new problem.

After the creation of the new alternatives and often after the acquisition of ready-made alternatives from the environment, an alternative is selected which seems to be the most appropriate to solve the problem (Mantzavinos 2001, ch. 4). Like all other solutions, this is an entirely *conjectural* one, which cannot guarantee success. Failure or success is determined when the chosen alternative is tried out in the environment. If the trial proves to be unsuccessful, then a new problem arises, and a choice process is triggered. If the chosen alternative proves to be successful, then it will be reinforced, and the next time the same problem arises, this solution will be reapplied. After it has been employed successfully many times, the solution will be standardized and will become a routine. In other words, the mind will avail itself of a ready-made solution to the problem, and thus every time it is confronted with it, it will be classified as an old problem and dealt with unconsciously. The general problem-solving framework and this never-ending process of the routinization of problem solutions is summarized in Figure 10.

8.3   THE EXPLAINER AS A PROBLEM-SOLVER

Explanatory activity can be analysed as a special kind of problem solving, in which the goal state is a target to be explained. The process in which a problem-solver is engaged in targeting an explanation can be conceptualized with the help of a rule-based account. There are a variety of ways to obtain the goal state, depending on the different forms that the explananda can take. In those cases, mainly in the sciences, that the explanandum is relatively well-specified and a formal deduction is possible, the explanation in the

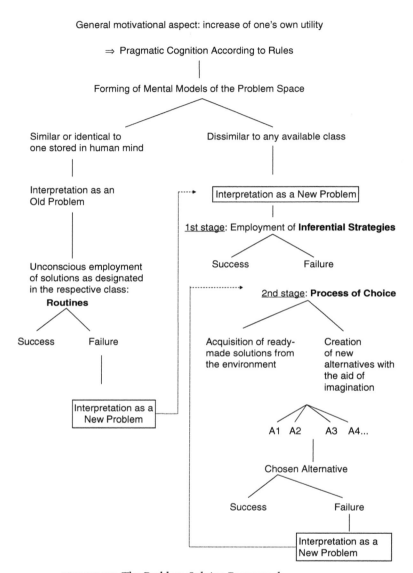

FIGURE 10 The Problem-Solving Framework

form of a deductive argument can be accounted for by a rule-based system. Thagard (2012, pp. 27ff.) uses the simple example:

> Anyone with influenza has fever, aches, and cough.
> You have influenza.
> So, you have fever, aches, and cough.

In this case the explanatory target follows deductively from the explaining propositions. In the rule-based approach, this example can be expressed by a rule like:

> If X has influenza, then X has fever, cough, and aches.
> Paul has influenza.
>
> _____
>
> Paul has fever, cough, and aches.

It is modus ponens that provides the connection between the rule and the explanandum; and in more complex cases this connection would be obtained by means of a sequence of applications of modus ponens as multiple rules are applied.[14]

In those cases in science, but most importantly also in everyday life, where the connection between the explainer and the target is not as strict as in a logical deduction, other kinds of inferences must be highlighted in order to obtain an accurate descriptive account of the explanatory practice. Analogy is an inferential strategy that is very frequently used by explainers. For example, the "process of induction model" (PI model) developed by Thagard (1988) has been applied to the wave theory of sound: it simulated the discovery of the wave theory of sound by analogy to water waves. But analogy is a prevalent inferential strategy employed to provide the mythical and religious explanations of cosmogony discussed in the last chapter, to help obtain mathematical explanations, etc.[15] The evocation of an analogy depends on the

---

[14] As Thagard (2012, p. 29f.) remarks: "In contrast to Hempel's account in which an explanation is a static argument, rule-based explanation is usually a dynamic process involving application of multiple rules."

[15] On analogical transfer see Holland *et al.* (1986, ch. 10.2).

precise structure of the mechanisms of both attention and recognition and on the contents of long-term memory and their organization.[16]

Newton's analogy ("universal gravitation") between projectiles and the moon and Darwin's analogy ("natural selection") between selective breeding and reproduction are further famous examples. And Nersessian (1992) presents a convincing case study of how Maxwell, building on the imagistic representation of the "lines of force" surrounding a bar magnet and of the "interconnectedness of electric currents and magnetic forces" that Faraday had produced, managed to provide a unified mathematical representation of the propagation of electric and magnetic forces with a time delay, using what he called a "physical analogy". In all of the mentioned cases, it is neither a deductive nor an inductive inference of the kind that philosophers of science usually analyse that does the inferential work, but analogies.[17]

[16]   See Simon (2000, p. 51f.): "Usually the explanation is finessed by pointing to similarities between two situations. Thus Bohr compared the hydrogen atom to the solar system (as Nicholson and other had already done) because they both involved a large central mass with one or small masses revolving about it and with the force field of the central mass connecting them. However, the matching process was apparently rather arbitrary and ignored important details: in particular, that the orbital electron would radiate energy, thus causing it gradually to collapse into the nucleus. Bohr then evoked Planck's quantum to 'explain' why the orbit could not change gradually hence could only jump from one discrete state to another. The choice of what to match and what not to match appears quite ad hoc, justified only by its success in explaining the atom's radiation spectrum. Thus, while there is not convincing explanation of why Bohr used this particular analogy in this particular form and in combination with the quantum hypothesis at this particular time, there is a quite direct explanation of why the analogy, so used, would lead him to an explanation of the wavelengths in the hydrogen spectrum that had been observed thirty years earlier and provided with a descriptive theory by Balmer's curve fitting. There is also a simple explanation of why no one did it much earlier: the prerequisite knowledge of the structure of the atom only became known through Rutherford's work in 1911."

[17]   The subjects in the study of Gentner and Gentner (1983) constructed a mental model of electricity in terms of either a water-flow analogy or the moving-crowd analogy; and specific inferences, which were sometimes erroneous, could be traced directly to the analogy. As Nersessian correctly points out (1992, p. 20) "[t]his result gives support to the view that analogies are not 'merely' guides to thinking, with logical inferencing actually solving the problem, but *analogies themselves do the inferential work and generate the problem solution*". Besides analogies, Nersessian (1992) discusses the role of further problem-solving strategies such as imagistic reasoning, thought experiment and limit case analysis.

Stressing the cognitive processes that take place when explainers solve a problem shifts attention from the products of the explanatory activity on which the traditional theory has focused to the process of explanatory activity itself. This does not deny that an analysis of explanatory arguments in terms of propositions and their normative appraisal is impossible or useless. The contrary is the case, as I will discuss in more detail in Chapter 11. But a general theory of explanation encompassing the whole range of explanatory activities – including the scientific ones – should pay the due attention to the broad range of cognitive process which, embedded in the respective institutional context, yield explanations rather than concentrate exclusively on (logical) deductive and inductive inferences (or on other kinds of inferences, like the inference to the best explanation[18]). One important aspect of these processes, as I have tried to argue here, is that situations represented by mental models rather than by an argument structured in terms of propositions might facilitate the drawing of inferences, since extensive computation might not always be needed. In a great number of everyday explanations where immediate representations of the situations are formed, the inferences can be quasi immediate.

To put it in yet other terms, *reasoning comprises more than algorithms*; and heuristics such as analogy play a decisive role.[19] This is the case not only in everyday explanatory practices but also in scientific explanatory reasoning. The problem-solving framework defended here can account for the mental representations that people form and the great variety of inferences they draw when explaining phenomena;

---

[18]   See the classic expositions of Harman (1965) and Lipton (2004).

[19]   See Nersessian (1992, p. 35). See also her comment (p. 9): "One value of having a mental-models form of representation is that it can do considerable inferential work without the person having to actually compute inferences and can also narrow the scope of possible inferences. For example, moving an object immediately changes all of its spatial relationships and makes only specific ones possible. The hypothesis that we do such inferencing via mental models gains plausibility when we consider that, as biological organisms, we have had to adapt to a changing environment. In fact, artificial intelligence researchers have run into considerable problems handling the widespread effects of even small changes in knowledge-representational systems that are represented propositionally."

and most importantly it can consistently elucidate the existing continuum between quite primitive and highly sophisticated explanatory practices. As will be shown in more detail in the next chapter, the rules of representation and rules of inference that have developed over the course of the history of science are very sophisticated and refined outgrowths of those used in everyday life and are due to the specific institutional context that has come to prevail in modern science rather than to the specific differences in cognitive mechanisms.

Work done by developmental psychologists seems to further support this view. The ability to provide explanations does not appear before a child's third year of life; and it is very rudimentary. A five-year-old child will provide as an answer to a why-question or how-question a phrase starting with because and followed by a repetition or paraphrase of the question (Keil and Wilson 2000, p. 3). A lot of findings, however, support the twin ideas of the "child as a scientist" (Carey 1985) and of the "scientist as child" (Gopnik 1996; Gopnik and Meltzhoff 1998). Empirical research suggests that very young children's learning and thinking are strikingly similar to much learning and thinking in science (Gopnik 2012). The similarity concerns first the conceptual structures which in both children and scientists are supposed to be theories; second the supposition that these theories provide explanations of events in the world; third that children and scientists both seem to be sensitive to evidence.[20]

---

[20]   The last point seems particularly striking, but Samarapungavan (1992) has carried out a direct test of this issue, presenting children with two theories, one of which was supported by empirical evidence and the other which was not. Children (even those in the first grade) showed a strong preference for the explanation that was empirically supported. This finding is important in the light of further findings that suggest that even if empirical support seems to play a big role in the explanatory activities of children, this does not go hand in hand with a preference for testable explanations. Brewer, Chinn and Samarapungavan (2000) argue that children, unlike (many modern) scientists, do not require explanations to be testable. This does not seem to be paradoxical, since many untestable explanations can account for all relevant empirical evidence – for example it was willed by a Supreme Being – and thus might appear attractive to children who have a preference for empirical support.

Gopnik (2000, p. 300) argues – quite boldly – that "explanation is to theory formation as orgasm is to reproduction – the phenomenological mark of the fulfilment of an evolutionary determined drive. From our phenomenological point of view, it may

Besides, it is important to stress that explanation targets are often represented nonverbally, so that explanatory activities often engage with *practical* problem solving. In medicine, for example, doctors employ visual hypotheses, say about the shape and location of a tumour, to explain observations that can be represented using smell, touch, sight, as well as words (Thagard 2012, p. 37). The problem solutions provided for these kinds of explanatory problems give rise to specific explanatory skills, that is, to procedural knowledge. *One should, thus, abandon the view that explanatory activity is exclusively a reasoning process and explanation the outcome of strictly theoretical reasoning.*[21]

Explanatory activities can thus be conceptualized as problem-solving activities which comprise the use of both declarative and procedural knowledge in the form of rules that the explainers have come to adopt in a trial-and-error process. Rules of representation,

seem to us that we construct and use theories to achieve explanation or that we have sex to achieve orgasm. From an evolutionary point of view, however, the relation is reversed, we experience orgasms and explanations to ensure that we make babies and theories." According to her suggestion there is a distinctive phenomenology of explanation which involves both the search for an explanation and the recognition that an explanation has been reached (p. 310f.): "We might call these experiences the 'hmm' and the 'aha'. In English, they seem to be expressed by 'Why?' and 'Because'. These experiences are obviously close to what we more broadly call 'curiosity' or 'interest', but they are not identical with them. We may engage in purely exploratory behaviours (the desire to open a locked cupboard, say, or climb a mountain, or see around a bend) that have no 'aha' at the end of them. Often they are connected to goal-directed or problem-solving behaviour, but they do not simply reflect the satisfaction that comes from achieving a goal. We may blunder our way to the exit, or use trial and error to find the right key to delete the file, and be happy we have done so, but without any 'aha'. Conversely, we may experience a splendid moment of illumination when we realize just exactly why it is thoroughly impossible to get what we want." Though these remarks are certainly valuable and important, I think that they overstress explanatory activity as a theoretical enterprise. As I argue in the text, however, there is also a component of procedural knowledge that must be paid attention to in order to obtain a complete account of explanatory activity.

21   As Keil and Wilson (2000, p. 3) correctly point out, "[a]s adults, we are often able to grasp explanations without being able to provide them for others. We can hear a complex explanation of a particular phenomenon, be convinced we know how it works, and yet be unable to repeat the explanation to another. Moreover, such failures to repeat the explanation do not seem merely to be a result of forgetting the details of the explanation. The same person who is unable to offer an explanation may easily recognize it when presented among a set of closely related ones."

rules of inference and rules of scope are learnt by individual explainers, are combined and lead them to provide explanations to the phenomena that interest them. *The issue is, thus, how explainers, individually or collectively, combine their cognitive abilities in the form of rules which they have by virtue of their biological make-up in interaction with their natural and social environment.* In this chapter, the focus was on the analysis of these cognitive abilities, abstracting from the natural and social environment. In the next chapter, we will complete the view by including the way that the feedback from the natural and social environment is integrated by the explainers.[22]

Summarizing, the problem-solving approach seems to provide a very useful accommodation of a great range of explanatory activities in science and everyday life. Additionally, it helps exemplify my thesis of the continuum between everyday, commonsensical knowledge and scientific knowledge as presented in Chapter 7. This thesis does not deny the difference in the quality of the scientific knowledge – in the case that is of interest here, of scientific *explanatory* knowledge – vis-à-vis other kinds of knowledge. But instead of presenting clear-cut demarcation criteria of a syntactic nature, my focus lies on the specific way that knowledge is imposed for criticism. The problem-solving framework substantiates this claim further: it is the specific conditions and constraints under which the feedback from the environment is obtained every time that a problem solution is tried out, that determine in the end the specific quality of the

---

[22] See the similar point of Nersessian in a review article on mental modelling in conceptual change (2008b, p. 392): "[...] the problem-solving practices of scientists arise out of and are constrained by basic cognitive capacities exhibited also in mundane problem solving, though of course not from these alone. The normally functioning human cognitive apparatus is capable of mental modeling, analogy making, abstraction, visualization, and simulative imagining. The sciences, through individual and collective efforts, have bootstrapped their way from these basic capabilities to the current state of play through consciously reflective development of methods of investigation aimed at gaining specific kinds of understanding and insight into nature, such as quantitative understanding. Of course, the development of these methods has been and continues to be a complex interaction among humans and the natural and socio-cultural worlds in which they are embedded."

hypothetical problem solutions. These specific conditions and constraints are different in "cognition in the wild" and in the tight institutional structure characteristic of modern science. It is the difference of the conditions of *criticism*, which give rise to the different qualities of accumulated problem solutions. And it is due to the long historical development of the institutional framework of modern science that these conditions have obtained and enabled the constant elaboration and refinement of problem solutions – in the case of interest here, of the scientific explanatory practices.

9 Explanatory Rules as Shared Rules

The emergence and change of the rules of an explanatory game is a complex process of a spontaneous interaction between the participants of the game, including several stages. The task is to provide an account of this social process by starting with the individual problem-solver, as conceptualized in the last chapter, and proceeding with the analysis of the social interaction between individuals as they are engaging in explanatory activities. The point of departure is the continuity thesis.

9.1 THE CONTINUITY THESIS

The continuity thesis has two dimensions. The first dimension concerns the micro-level: the innate cognitive structure of the mind does not vary among hunter-gatherers and scientists; in other words, there is no fundamental difference in the cognitive architecture of the members of our species.[1] The second concerns a straightforward macro-dimension: there is a continuum between everyday, common-sensical knowledge and scientific knowledge in the sense that scientific knowledge is the outcome of the constant elaboration and refinement of everyday, commonsensical knowledge.

I provided a specific conceptualization of the first dimension of the continuity thesis in the last chapter in terms of the problem-solving framework. Although I have offered a rather general account of problem-solving, meant to accommodate a number of issues and phenomena important for my general project – most importantly the

---

[1]   See the very interesting discussion by Mithen (2002). Simon (1992, p. 105f.) puts this issue as follows: "The scientist does not think in ways that are qualitatively different from the ways in which other professionals think, or the ways in which college students think when confronted with puzzles to solve in the psychological laboratory, or the ways in which T.C.Mits (the common man in the street) thinks."

routinization of problem solution activities – there are, of course, a series of limitations. The most important is perhaps the rather vague stance towards the question of whether the innate basis of cognition should be viewed with the premises of a domain-general account or rather with the premises of a domain-specific, more or less modular account.[2] Though there is considerable debate in cognitive science about this issue, I think that it is legitimate to remain agnostic on this, since the core of my approach will remain unaffected regardless of which of those two views is adopted.

An integral part of the micro-level dimension of the continuity thesis is the developmental account of the "child as a scientist" or the "scientist as a child" that I discussed in the last chapter. But here, again, it is appropriate to point at some limitations. Though an analysis of the cognitive development from an infant to the adult does provide important insights into and support for the continuity thesis, it tends to neglect the external resources by which scientific and here, more specifically, explanatory activity needs to be supported. It is an intricate institutional context that enables participants to the scientific explanatory game to play it in such a way that outcomes are produced that we tend to evaluate positively using an array of different criteria. The developmental line of research appears to consistently neglect its importance;[3] and this seems to be a serious shortcoming. In order to have a complete view, which will also be inclusive of the macro-dimension, the social context is to be incorporated.

[2]   The domain-general account is defended by Simon and a series of other authors, whereas the modularity view has been prominently defended by Fodor (1983), Cosmides and Tooby (1992), and Gigerenzer (2008). For an excellent discussion with respect to the continuity thesis, see Carruthers (2002).

[3]   This might be the case because, as Carruthers argues (2002, p. 84), developmental psychologists tend to oppose "social constructivist" accounts of science, which claim that wider social and political forces give rise to scientific change. However, endorsing the importance of the institutional context is a far cry from endorsing such a view. As Carruthers correctly points out (2002, p. 84): "Scientists do not, and never have worked alone, but constantly engage in discussion, co-operation and mutual criticism with peers. If there is one thing which we have learned over the last thirty years of historically oriented studies of science, it is that the positivist-empiricist image of the lone investigator, gathering all data and constructing and testing hypotheses by himself or herself, is a highly misleading abstraction."

9.2   SHARED RULES AS SHARED PROBLEM SOLUTIONS

As we have seen in Chapter 8, when an individual faces a new problem, he always possesses one possible way to solve it, that is, to utilize ready-made solutions from the environment. Such direct learning, which consists in the adoption of ready-made solutions rather than the creation of new alternatives ex nihilo each time a new problem arises, goes hand in hand with the emergence of a process of communication. The result of the communication is the adoption of problem solutions introduced into the individual cognitive system from the cognitive systems of the other individuals. After the communication acts have taken place, both the sender and the receiver possess more or less similar cognitive rules by virtue of the fact that the content of the cognitive rules of the sender are to some degree adopted by a receiver. Acts of communication then lead to the adoption of shared rules by individuals and give rise to correspondent *shared mental models* (Denzau and North 1994, Mantzavinos 2001).

Direct individual learning based only on direct feedback that a problem-solver gets from his natural and social environment is an immensely slow process. It is rather indirect learning from other individuals that contributes most to the growth of individual knowledge. Only when agents can share their (experiences or) problem solutions with other agents can their knowledge grow rapidly. Some simulation models support the thesis that the rate of direct learning is much slower than the corresponding rate of culturally or socially mediated learning.[4]

Cognitive anthropology has developed a number of methods to explore empirically the content of shared mental models, or more generally, cognitive structures across cultures (D'Andrade 1995). This branch of research (which abandons both the structuralist and interpretive views of culture), besides being more testable, is surely on the right track by virtue of its insistence on rejecting the view of

---

[4]   Hutchins and Hazlehurst show, for example, that cultural learning halves the time required to learn the relation between moon phases and tides (1992, p. 703).

culture as a complex, integrated whole and its suggested view of
culture as "socially inherited solutions to life problems" (D'Andrade
1995, p. 249). This is particularly important for our discussion, since
the term "culture" is an extremely vague concept, prone to many
misunderstandings;[5] and it is better not to use it in the present context
and to speak instead of shared problem solutions, which, according to
my problem-solving framework model, can be best conceptualized as
shared rules.[6]

The emergence of those shared problem solutions often does not
involve any kind of conflict between the interacting problem-solvers.
In many other cases, however, such an emergence takes place under
non-cooperative settings. Where conflict is involved, and depending
on the exact characteristics of the situation, different kinds of social
rules emerge in the process of social interaction that provide solutions
to these problems. These rules are of a normative nature, regulating
the respective conflict at hand. Depending on the agency that enforces
the normative rules, a useful distinction is between formal rules and
informal rules. The formal social rules or *formal institutions* are
enforced by the state. The informal social rules or *informal institu-*
tions are enforced by the other members of the group.
The mechanisms of emergence, change, and adoption are different
for the different types of rules (Mantzavinos 2001). What is mostly
important for our purposes here are the characteristics of the process
of emergence, change, and adoption of social norms, so I will briefly
discuss them in the next section, before turning in the final section to
the application of this analysis to the specifics of explanatory norms.

---

[5]   See D'Andrade (1995, p. 250): "Of course, one can study in a scientific way the elements
of culture, but to find out why cultural elements exist and how they fit together one has
to step outside the concept of culture and look for whatever it is that creates and
organizes these elements, such as the problem of biological reproduction, or the problem of getting food out of the environment, or human cognitive limitations, or
personality needs, or whatever."

[6]   As Denzau and North (1994, p. 15) correctly observe: "The cultural heritage provides
a means of reducing the divergence in the mental models that people in a society have
and also constitutes a means for the intergenerational transfer of unifying perceptions.
We may think of culture as encapsulating the experiences of past generations of any
particular group."

9.3 THE NORMATIVE DIMENSION: THE INSTITUTIONAL FRAMEWORK

Social norms exist because they provide solutions to different social problems in which conflicting individual interests prevail.[7] Social norms vary tremendously among different places and times; therefore it is nearly impossible to discern stylized problems to which social norms provide an answer. "A norm exists in a given social setting to the extent that individuals usually act in a certain way and are often punished when seen not to be acting this way" (Axelrod 1986, p. 1097).

How social norms emerge is an issue of fundamental importance. The rational choice account of norms, originally and most influentially presented by James Coleman (1990a, 1990b), and developed further by a series of authors, is unable to describe the process in a consistent way: rational agents, who are assumed to be forward-looking, will sanction a deviator only when their immediate benefit outweighs their sanctioning costs. However, in a rational choice framework this is by no means evident, mainly because rational agents will have an incentive to cheat and let the other members of the group sanction the deviator. The way out of this "second-order free-rider problem" cannot consistently come from a rational choice perspective in the case of social norms that evidently emerge without any commitment device on the part of the agents. Coleman (1990b, p. 53) gives an inherently inconsistent solution to this problem when he writes: "The existence of a norm facilitates achievement of the social optimum by making use of the *social relationships* that exist in a social system to overcome the second-order free-rider problem." But where do these social relationships come from? In a consistent application of the rational choice framework these social relationships cannot provide a solution to the second-order free-rider problem, but only transform it into a third-order public goods problem in which every individual in a social relationship has an incentive to cheat, and so on, ad infinitum.

---

[7]   The discussion in this chapter is based on Mantzavinos (2001, sec. 7.3).

An alternative is to provide an account of the spontaneous emergence of social norms in an evolutionary process of the invisible-hand type,[8] using our problem-solving framework. This allows agents to be motivated by an increase in utility, but mainly focuses on their ability to learn by trial and error, which gives rise to a routinization of their behaviour. There are three main stages to this process.

In the *first stage*, when agents are confronted with a situation where social conflict is involved with which they have had little or no experience, that is, with a new problem situation, they "will call on the experience they have of similar situations they have faced in the past in deciding what action to take" (Binmore and Samuelson 1994, p. 51). They will, in other words, employ some *inferential strategies*, very frequently analogies, in order to reach a solution to the new problem. Thus, they will be aided by their experience of analogous situations to solve the problem at hand. If this does not lead to the solution of the problem, then they will create new alternative solutions and choose the one from which they expect the highest increase in utility. If the trial is successful, they will tend to reapply the solution again in the future every time that they find themselves in similar situations until it becomes a routine. Thus, an adaptation to the new circumstances in the environment will have taken place, and a regularity in individual behaviour will be observed.

This behaviour, if successful in the sense of increasing utility for the agent who employs it, will be imitated – and this is the *second stage* of analysis – by other individuals who also expect a utility increase. Research in social psychology has shown that successful models of individual behaviour tend to be imitated in a process of observational learning (Bandura 1986).[9] Learning by imitation leads to a diffusion of the behaviour of the innovator and thus to many or all members of the group following the same rule. This diffusion of the

---

8   For a detailed characterization of such a process, see Ullmann-Margalit (1978) and Nozick (1994).

9   See also Opp (1982, p. 29): "If, for example, a subset of members of a group performs actions which are rewarded and are observed by other members, most of the observers will expect to receive the same or similar rewards if they imitate the models."

behaviour among the members of the group involves the diffusion of shared solutions to the specific problem of social conflict: it concerns the knowledge of how to behave in a specific context and, thus, the routinized way to solve a specific problem in a specific context of social interaction. (The individuals involved need not always be conscious that the behaviour that they engage at the same time offers a solution to a problem that, viewed from an observer's third-party perspective, is classified as a social one, although often they will be.)

The *third stage* concerns the analysis of the sanctioning of the deviators by those who follow a certain behaviour. Although those who punish the deviators do so in order to increase their own utility, this does not mean that members of the group of punishers engage in cost-benefit calculations about whether it is beneficial or not for them to punish the deviator. It is moreover the case that they tend to sanction a deviation every time it occurs, because they generally hold that such a deviation is "wrong", that is, rejectable. A non-exhaustive set of general factors that can provide an account of this sanctioning stance towards deviators includes (a) the monitoring of social norms is often intrinsic to the social relationship, therefore enforcement occurs as a by-product of social interaction;[10] (b) those who conform usually attain higher status and power within the group, and they are therefore indirectly rewarded for their enforcing activity;[11] (c) probably the most important factor has to do with the feeling of satisfaction that human beings have when they live in a predictable environment, so that whenever their expectations regarding the behaviour of others are disappointed, they tend to have a feeling of uneasiness. In other words, the mere fact that an expectation is disappointed brings about a feeling of uneasiness (Sugden 1998,

---

[10]   On this point see Nee (1998, p. 87), Nee and Ingram (1998, p. 28) and Nee and Strang (1998, p. 709).

[11]   McAdams (1997, p. 365) stresses the role of esteem in the emergence of social norms, arguing that "[t]he key feature of esteem is that individuals do not always bear a cost by granting different levels of esteem to others. Because the cost is often zero, esteem sanctions are not necessarily subject to the second-order collective action problem that makes the explanation of norms difficult." Costless esteem is therefore a by-product of social interaction. See also Nee and Ingram (1998, p. 30).

p. 85), vengefulness, or "moralistic aggression" (Trivers 1971).[12] Summarizing, due to these three factors every deviation from a standard behaviour tends to be sanctioned. The conditions for norm emergence are therewith fulfilled, that is, a behaviour is followed by many members of a group *and* deviation is sanctioned by group members.

Coming to the issue of the stability and change of social norms, Axelrod (1986) has shown that in order for a norm to be stable in a population, an important degree of vengefulness must exist among the members of the population. In this context, for stability to be secured a metanorm is required according to which one must punish those who do not punish a defection. Thus, the persistence of a norm depends to a great degree on the development of such higher-level metanorms, and it is the specific characteristics of the concrete historical situation which must be analysed in order to individuate the existence of such metanorms.[13]

However, the existence of sanctions does not imply that social norms can never change. A change in norms presupposes, according to the problem-solving framework, *a change in the environment that calls for a new adaptation.* Such a change will either induce a subsequent change in the incentives to sanction the deviating behaviour or it will alter the structure of the social problem. This in turn means that an innovative agent will try out different problem solutions and then adopt a routine that will subsequently be imitated by other agents and in the end his social behaviour will become – under the right conditions – a new norm. Norms are adaptive devices to environmental changes in the social setting and, thus, they are usually

[12] See the remark of Opp (1982, p. 67): "Even in the case of umbrellas in the rain, people who strongly prefer to use umbrellas themselves would also prefer others to do so too. People do, in fact, generally prefer to be able to predict their environment. Thus we may assume a positive correlation between the intensity of preference for a behaviour and the degree to which irregular behaviour of others is costly."

[13] But Elster (1989a, p. 133) remarks that it is improbable that such metanorms, which sanction people who fail to sanction people who fail to sanction people, and so on, might exist in the real world. "Sanctions tend to run out of steam at two or three removes from the original violation" (Elster 1989b, p. 105 n.4).

transformed whenever a relatively permanent shift in the environment takes place. Nevertheless, from an observer's point of view, this environmental change need not be an objective one in order for an adaptational change of norms to be triggered. It suffices if the pioneering individual perceives his given environmental situation as a different one and tries out a new behaviour, and the rest follow him because they expect that this new behaviour will also increase their own utility.

The problem-solving framework which highlights the cognitive-emotional aspects of norm emergence and change avoids many difficulties of the rational choice approach. The standard weakness of the latter approach is that a social norm is viewed as an equilibrium outcome, without a specification of the factors that lead to a particular equilibrium from the array of possible multiple equilibria. This weakness is remedied by my approach, presented briefly here. Besides, my account does not slide into the standard sociological view, which can be best summarized in the following twin theses: "The thesis of the normative regulation of behaviour", which states that if norms are institutionalized and internalized, then the behaviour that is in accordance with the norms occurs; and "the transmission thesis", according to which social norms, by their bare existence in a society, become well known to its individuals and become internalized quasi-automatically because of fear of sanctions.[14] This standard sociological view is defective, because it cannot account for the simple fact that *self-interested* people need not pay any attention to norms and they, in fact, very often do not. On my account, on the contrary, mental models will be formed by deviators, specifying which acts are accompanied by which sanctions; and only in this sense will an internalization of norms, so often employed in sociology, take place. Such an acquisition of certain patterns of behaviour on the part of individuals, once adopted, is bound to appear to an external observer as a blind, norm-driven behaviour. The fact that these patterns of behaviour may

---

[14]   See more details in Opp (1979, pp. 777ff.).

often change is enough to prove that what seems to be a very rigid behavioural pattern is only conditioned on the environmental feedback and the learning histories of the individuals.

My account has, of course, a series of limitations. However, my main intention in this section was not to develop a full-blown account of the emergence, change, and adoption of social norms and other informal institutions, but to show that the problem-solving framework can consistently and sufficiently accommodate the normative dimension. A great array of situations encountered in the environment by individuals is perceived through normative filters (not necessarily by *a single* normative filter), and a great range of problem-solving activities of the individuals is fundamentally guided by norms. Though I have focused here on informal institutions, more specifically social norms, the formal institutions, that is, the normative rules enforced by the state, are also a constitutive part of the institutional matrix.[15] In a nutshell, problem solving always takes place within an institutional framework that cannot be assumed to be "given", but that has emerged in a historical process of individual and collective trial and error; and any institutional framework is only behaviourally relevant because it is anchored in the minds of the people.

9.4   ON THE EMERGENCE, CHANGE AND ADOPTION OF EXPLANATORY RULES

It is time to pull all the strings together and apply the insights of this and the last chapter to the issue that interests us here. The rules of the explanatory games emerge and change in a process of spontaneous interaction between the participants of the game. The analysis of explanatory activity, conceptualized as problem-solving activity, and of the explainer as a problem-solver was the first step undertaken in the last chapter. In this chapter we have provided a general account of the emergence, change, and adoption of normative rules in an

---

[15]   For a full-blown account of the emergence, change, and adoption of formal institutions, see Mantzavinos (2001, ch. 8).

invisible-hand process of social interaction. The upshot of this discussion is that normativity emerges in social groups when shared behavioral regularities anchored as routinized problem solutions in the minds of the members of the group are diffused and the deviators sufficiently sanctioned. All of these shared behavioral regularities always start as individual regularities.

I claim that explanatory rules avail of this quality of normativity. Explanatory activities are norm-guided activities in that the explainers adopt rules of representation, rules of inference and rules of scope which aim at *providing adequate solutions* to the explanatory problems at hand. The participants to a specific explanatory game have adopted the rules over time; these rules incorporate the normative standards that guide the processes of the discovery and the justification of explanations as well as the modes of their communication, dissemination and adoption.

It should be obvious that all explanatory games which are constituted by normative rules that the participants of those games follow are themselves embedded in broader normative networks and normative matrixes: the explainers are also participating in myriad other social contexts where different normative rules prevail and which they sometimes adopt. These normative frameworks exert – naturally – an influence on the rules of the explanatory games. It is a whole institutional matrix of informal and formal institutions, thus, which constrains the explanatory games, regulating to a greater or lesser degree the problem-solving activities of the explainers.

In Chapter 6 I provided an abstract characterization of an explanatory game in terms of its constitutive rules and the rules of representation, the rules of inference and the rules of scope that the participants follow. If one wants to describe the different kinds of games, then the precise content of the rules must be specified, of course. The emergence and change of the rules of the respective game takes place in a process of learning of the explainers who themselves engage in problem-solving activities. As I have argued *in extenso* in Chapter 8, this learning process fundamentally depends on the nature of the feedback that the explainers receive from their environment. It is the environmental feedback which leads either to the routinization and adoption of a problem solution or to the rejection and the search for alternatives. The simplistic abstraction of an explainer figuring out alone the specifics of the book of nature is misleading. The situation is much more complex in that the explainer receives feedback from both the natural and the social environments, which is judged within the normative dimension in which he has emerged *peu à peu* in a long historical process. It is the analysis of the conditions of the specific environmental feedback which provides the key to understanding the mechanics of the rule emergence and change. More specifically, I want to claim that it is the specifics of the way that the problem solutions tried out in the environment of the explainers are imposed for *criticism* that one should focus on. It is the specific conditions and constraints under which the feedback from the environment is obtained every time that a problem solution is tried out that determine the quality of the hypothetical problem solutions. These specific conditions and constraints differ from case to case. I will briefly mention the case of children, the feedback that they tend to receive from their environment and the games that they come to play. I will then also mention the case of religious explanatory rules, before turning to scientific explanations and how they are shaped by the tight institutional structure of modern science.

(a) **Children in explanatory games**. When children are involved in explanatory games, they typically find themselves in interaction with adults as they provide their tentative solutions to explanatory problems. The feedback that they receive is filtered by the set of norms that they have adopted from their immediate family environment; and they (most importantly) accept the authority of the adults when evaluating their feedback. They come to accept as satisfactory all those explanations that their family environment provides them with, and though their critical faculties are exerted, this is done only to a rudimentary degree. The evidence provided to the children in their early years, but also in later years, in the form of the feedback from the environment, takes a back seat to *authority* (Faucher *et al*. 2002, p. 351). When a young boy in North Korea asks his mother why the portrait of Kim Il-sung hangs in their living room, he accepts the explanation that it is there because he is the "Great Leader" without further ado, because it comes from his mother. When children are taught the basics of the heliocentric theory of the solar system, the challenge is to make them understand the theory itself. They do not assess the evidence for and against it, and in fact they do not even ask for any evidence.[16]

The rules of representation, rules of inference and rules of scope that children *come to adopt* while playing their explanatory games are quite elementary. But they tend to change according to the environmental feedback that they receive. A concrete example can exemplify the point better. A substantial body of cross-cultural research reviewed by Vosniadou *et al*. (2008) suggests that during the preschool years children construct initial representations of the earth based on the feedback of everyday experience in the context of lay culture. According to this initial concept, the earth is flat, stable, stationary, and supports physical objects. The "laws of up/down gravity" are the rules of inference that govern the objects located on the earth, and space is organized in terms of the dimensions of up and down. The sky and solar objects are located above the top of the flat earth and the rules of scope specify that the explanations apply to a geocentric universe.[17] In the social context of the elementary school, children are exposed to an authoritative criticism by their teachers, and the feedback that they receive violates practically the whole range of problem solutions that they have come to adopt. "New problems" arise that call for new representations and inferences. The feedback that they receive suggests the current scientific view according to which the earth is a planet, that is, an unsupported, spherical,

[16]   This is similar to the case of students in the university and the way they learn scientific theories: if the university professor teaches it and seems to endorse it and the textbook presentation is clear enough, then this is typically sufficient for the students to accept it. But we will come to this point later.

[17]   For more details on that, see Vosniadou and Brewer (1992, 1994).

astronomical object, which rotates around its axis and revolves around the sun in a heliocentric solar system.

Laboratory experiments show that children come to change their rules of representation very slowly and to build alternative mental models of the earth, *starting from and revising the initial one* of a flat earth until they adopt the representation of the earth as a sphere. The younger children tended to represent the earth as a square, rectangle or as a disc-like flat, physical object; some other children came to adopt the representation of a dual-earth, according to each there are two earths: a flat one on which people live and a spherical one which is up in the sky and which is a planet; yet other children have formed representations of the earth as a hollow sphere, according to which the earth is spherical but hollow inside; yet other children's representations have changed such that they have adopted the model of earth as a flattened sphere or truncated sphere with people living on its flat top, covered by the dome of the sky above its top.

Based on the respective representations children also have come to *revise their respective rules of inference*, which allow them, for example, to make claims about unobservable properties of the earth, that is, that it is supported or that it has an end. Those who have come to adopt the model of the hollow sphere use the rule of inference according to which people live on flat ground inside the earth because they would fall "down" if they lived on the surface of the spherical earth. Similarly, those children who have adopted the view that the earth is a flattened or truncated sphere also infer that people live on flat ground above the top of the earth. Adopting the new explanation encapsulated in the current scientific view of the matter involves, thus, a change in the children's rules of representation and rules of inference. This takes time and intellectual effort and less emotional resistance than in alternative cases where adults are involved simply because the authority of the teachers makes children concentrate on the change of their rules such that they become more consistent with the feedback that they receive from their teachers (and not with the application of specific rules to evaluate any kind of evidence).

Cultural artefacts like maps and globes can be internalized and used in the process of revising mental representations that are based on everyday experience. As Vosniadou *et al.* (2008, p. 25) note, "[o]ur studies of children's reasoning in astronomy provide important although preliminary information about how individuals can construct mental representations that are neither copies of external reality nor copies of external artefacts, but creative synthetic combinations of both. This suggests that the cognitive system is flexible and capable of utilizing a variety of external and internal representations to adapt to the needs of the situation."

In a nutshell, though children do get feedback from their natural and social environment in their explanatory activities and the explanations that they provide do benefit from this feedback, the centrality and the difficulty of systematically pursuing, producing and appraising empirical evidence prevalent in other games – most importantly, as we will see later, in scientific explanatory games – is simply not given in this case.[18] The *criticism* is mostly directed towards the reorganization of the rules of representation and the rules of inference, so that a successful problem solution to the new problem which they encounter can be produced. The *criticism* is directed in this way because of the prevalence of the *authoritarian normative dimension* in which children are forced to operate.[19]

---

[18]   For more on this point, see the discussion by McCauley (2000, p. 66f.).

[19]   The position that I have articulated in the main text provides, I think, also a satisfactory answer to "the 1492 problem" which was at the centre of the controversy between Gopnik (1996b) and Giere (1996). To the already discussed opinion of Gopnik, which amounts to the view that "science is a kind of sprandel, an epiphenomenon of childhood" (Gopnik 1996a, p. 490), Giere opposed that science as we know it did not exist in 1492 (Giere 1996, p. 539): since the cognitive devices that give rise to science already existed in the Pleistocene, the question emerges why humans have been doing science only for the last 500 years. Gopnik's reply focuses on the availability of evidence (Gopnik 1996b, p. 554): "My guess is that children, as well as ordinary adults, do not, in fact, systematically search for evidence that falsifies their hypotheses, though they do search for evidence that is relevant to their hypotheses and they do revise their theories when a sufficient amount of falsifying evidence is presented to them. In a very evidentially rich situation, the sort of situation in which children find themselves, there is no point in such a search; falsifying evidence will batter you over the head soon enough." The various social and historical factors that

(b) Rules of religious explanatory games. Turning to religious explanatory practices it is important to point to some various large-scale indications suggesting that many aspects of religious cognition rely far less on *specific* (informal) institutional foundations than is sometimes presumed (McCauley 2000, p. 73f.). First, religion dates from the prehistoric past and its birth is less exceptional than, for example, the birth of science. Second, many religious ideas have recurred throughout human history across a great variety of institutional settings. All religious systems, including Buddhism, tend to anthropomorphize the social and natural world and to proceed to explanations based on agents and their activities. Independently on the level of complexity of the representations, for example, in the form of the Holy Trinity, it is commonly some postulated superhuman *agent* who is endowed with unconventional powers. Third, religion occurs in every human culture, and even when a particular religion becomes extinct, religion itself does not disappear but re-emerges in other forms.[20]

she mentions focus on the unprecedented amounts of new evidence that became available with the dawn of modern science and the development of telescopes, microscopes, air pump, etc. "The contribution of technology also seems relatively straightforward. If you want evidence about the stars, then telescopes help a lot. The more subtle technology that allows a mathematician in Italy to know what an astronomer has seen in Denmark helps even more" (Gopnik 1996b, p. 554). The increase in the availability of leisure is also mentioned as a factor that has created an environment in which the cognitive mechanisms with which natural selection has endowed children have begun functioning actively in adulthood and could have given rise to modern science. As Faucher *et al.* (2002, p. 340) correctly point out, however, "[t]he problem, as [they] see it, is that just about all factors cited in Gopnik's explanation of the emergence of science in the West were present at that time – and indeed much earlier – in China". Giere (1996, p. 539) makes also the same point: "But to suppose that these capacities alone were sufficient to generate modern science would leave us with no explanation of why modern science did not arise, for example, in Greece in 492 BC or in China in 1492 BC."

The crucial point according to my account is to highlight *the possibility and varieties of criticism* when analysing the explanatory games and this is given in the prevailing normative framework as it is embedded in the matrix of the prevailing informal and formal rules.

[20]   As McCauley (2000, pp. 74ff.) correctly points out: "[N]either the birth nor the persistence of religion critically depends on any special cultural traditions. (If the experience of the twentieth century is representative, religions persist, as often as not, even in the face of direct suppression.) At least in comparison to interest in scientific ideas, the

The rules of representation that the players in the religious explanatory games come to adopt are, thus, centred on the attribution of agency to different phenomena. Such rules seem to be ubiquitous, and attributing agency to different parts of the physical universe seems to be nearly a cognitive compulsion in humans (Mithen 1996, pp. 164ff.).[21] The means of representation are typically narratives (and often theatre). They require extremely limited mnemonic resources, and they are characterized by their simplicity. But, of course, narratives are driven by agents – it is their actions that give the structure and sometimes perplexity to the plot of narratives. The (metaphysical) presuppositions of universal agency attribution constrain the mental representations of the problem situation of the explainer: only specific *narratives* can provide the adequate representation of a problem situation, narratives that describe and evaluate individual and collective actions.[22] The rules of inference that religious explainers adopt are also in a way parasitic on the attribution of agency (and the attribution of the respective causal powers) to various parts of the universe. A specific class of inferential strategies is quasi-naturally triggered given the mental representations used, and in narratives the conceptualization of deities as agents licenses inferences about their (sometimes extraordinary) powers, values, preferences, etc.[23] Rules of logic

appeal of religious ideas is in no small part a function of our cognitive predilections. [...] In contrast to science, religion relies far more fundamentally on our standard cognitive equipment.

"[...] Religion rests on far more basic cognitive abilities, the most important of which is the ability to distinguish agents and their actions from other things and events in the world. Agents are entities in our environments who merit very different treatment from everything else. Their detection is critical to human's physical and social survival, and research in developmental psychology [...] affirms that children possess this ability in their first year of life."

[21] See the remark of McCauley (2000, p. 77): "We not only see faces in the clouds; we routinely talk about our cars' and computers' recalcitrant moods. Advertisers have anthropomorphised everything from cleaning products to vegetables to airplanes."

[22] This point can best be seen in contrast to the mental representation of events, which do not include any kind of agency.

[23] I think that my account is largely consistent with the one suggested by Pascal Boyer (2000, p. 100): "The main point of [my] account is that religious concepts are to a large extent constrained by their connection to intuitive ontology, that is, a set of categories and inference-mechanisms that describe the broad categories of objects to be found in

encapsulating consistency requirements are not always respected, and inferences to extraordinary causal powers help solve the explanatory problems at hand. Finally, the rules of scope adopted by the players of the religious explanatory games usually allow the application of the explanations to a huge variety of phenomena, partly due to the rules of representation used and partly due to the lack of rigid consistency requirements and other requirements of the rules of inference.

It is important to stress the kind of normative framework that is in place and that filters the feedback that religious explainers get from their environment. It seems to be characteristic that though the norms regulating the acceptability of explanations are very liberal in a variety of ways (judged from the perspective of other explanatory games), the other norms that constitute the institutional framework are not. Participants in religious explanatory games have typically also adopted a great array of other social norms that filter the feedback that they get from their environment. The most important is certainly that the explanations and the rituals that most commonly accompany the narrower explanatory practices are taken to be handed down from one generation to the next *without change* and are not supposed to be radically criticized. Criticism, to be sure, is allowed, but to an extremely limited degree, mostly in the form of criticizing the comments on the explanations recorded in "holy scriptures". The sanctioning of the deviators is very often thorough and sometimes cruel. The organizational structures within which religious explanations are transmitted, that is, churches, provide additional forms of enforcement of the norms with the use of various means. Sometimes the rigidity of enforcement of the norms reaches the physical extinction of the deviators.

the world (PERSONS, ARTEFACTS, ANIMALS, PLANTS) and some causal properties of objects belonging to these categories. These early developed, universal categories and inference-mechanisms deliver intuitive expectations about the likely states of such objects and the likely explanations for their changes. Intuitive ontology includes those domain-specific structures known in the literature as 'naïve theories' of various domains: intuitive physics, intuitive biology, and theory of mind, for instance. [...] Religious concepts are constrained by intuitive ontology in two different ways: [1] they include explicit *violations* of intuitive expectations, and [2] they tacitly activate a *background* of non-violated 'default' expectations."

In a nutshell, even if the religious explanatory practices themselves seem to take place according to rules of representation, rules of inference and rules of scope which allow for the emergence of an immense variety of explanations, and the more narrow norms of appraising the quality of the explanations seem to be very liberal, it is mainly the rest of the normative institutional framework that restricts *criticism*. It is mainly the details of the commentaries of the documented form of the *explanations that are not supposed to change* that are subject to criticism. The *criticism* is directed in this way because of the prevalence of the *dogmatic normative dimension* in which religious explainers usually find themselves operating.

(c) Rules of scientific explanatory games. Karl Popper, in his classic essay on "Back to the Presocratics", has highlighted the role of the critical tradition in the emergence of modern science, which goes back to the innovations of Ancient Greek philosophy (1958/1989, pp. 148ff.):

> The early history of Greek philosophy, especially the history from Thales to Plato, is a splendid story: It is almost too good to be true. In every generation we find at least one new philosophy, one new cosmology of staggering originality and depth. How was this possible? Of course one cannot explain originality and genius. But one can try to throw some light on them. What was the secret of the ancients? I suggest that it was a *tradition – the tradition of critical discussion*.
>
> I will try to put the problem more sharply. In all or almost all civilizations we find something like religious and cosmological teaching, and in many societies we find schools. Now schools, especially primitive schools, all have, it appears, a characteristic structure and function. Far from being places of critical discussion they make it their task to impart a definite doctrine, and to preserve it, pure and unchanged. It is the task of a school to hand on the tradition, the doctrine of its founder, its first master, to the next generation, and to this end the most important thing is to keep the doctrine inviolate [. . .] In this way all changes of doctrine – if any – are

surreptitious changes. They are all presented as re-statements of the true sayings of the master, of his own words, his own meaning, his own intentions[...]

[T]he conjecture that Thales actively encouraged criticism in his pupils would explain the fact that the critical attitude towards the master's doctrine became part of the Ionian school tradition. I like to think that Thales was the first teacher who said to his pupils: "This is how I see things – how I believe that things are. Try to improve upon my teaching." (Those who believe that it is "unhistorical" to attribute this undogmatic attitude to Thales may again be reminded of the fact that only two generations later we find a similar attitude consciously and clearly formulated in the fragments of Xenophanes.) At any rate, there is the historical fact that the Ionian school was the first in which pupils criticized their masters, in one generation after the other. There can be little doubt that the Greek tradition of philosophical criticism had its main source in Ionia.

It was a momentous innovation. It meant a break with the dogmatic tradition which permits only *one* school doctrine, and the introduction in its place of a tradition that admits a *plurality* of doctrines which all try to approach the truth by means of critical discussion.

It thus leads, almost by necessity, to the realization that our attempts to see and to find the truth are not final, but open to improvement; that our knowledge, our doctrine is conjectural; that it consists of guesses, of hypotheses, rather than final and certain truths; and that criticism and critical discussion are our only means of getting nearer to the truth. It thus leads to the tradition of bold conjectures and of free criticism, the tradition which created the rational or scientific attitude, and with it our Western civilization, the only civilization which is based upon science (though of course not upon science alone).

Scientific explanatory activities unfold within the intricate institutional framework of modern science. It is a historical contingency that the informal and formal institutions of modern science have come to prevail in a long evolutionary process in the West. On the one hand, the *informal institutions* encapsulating the critical tradition coming from the Ancient Greek philosophy, weakened in the course of many centuries and revived during the Scientific Revolution; and on the other hand, the emergence of *competitive political structures* (Bernholz *et al.* 1998, Jones 2003) that has considerably increased individual freedom allowing at the same time criticism without pernicious consequences for the critic. In the modern era an intricate institutional matrix has come to prevail in most parts of the world, which has further cemented the freedom of expression that naturally gives rise to a plurality of opinions and fosters the competition among different views. The gradual historical development of the set of organizational structures characteristic of *modern universities* enabled moreover the pooling of a great array of resources – intellectual and material – dedicated to the constant generation and criticism of solutions to abstract theoretical and practical problems.

Scientific explanatory games are, thus, embedded in these broader normative structures of the modern world. What appears to distinguish them from the other explanatory games unfolding in ordinary life (including the religious ones) is the relative sophistication and systematicity by which empirical evidence is generated and assessed, something that is enabled by the social and cultural arrangements as they are encapsulated in the institutional framework of science. It is the possibility of criticism provided by this framework which acts as a corrective to the error-prone problem-solving activities in which scientists, like ordinary people, are engaged and to the explanatory activities with which we are primarily concerned here;[24]

---

[24] Longino (1990, p. 76) suggests that it is transformative criticism which yields the objectivity of science: "Objectivity, therefore, turns out to be a matter of degree. A method of inquiry is objective to the degree that it permits transformative criticism. Its objectivity consists not just in the inclusion of intersubjective criticism but in the degree to which both its procedures and its results are responsive to the kinds of criticism described. [...] Method must, therefore, be understood as a collection of social, rather than individual processes, so the issue is the extent to which

errors ranging from fallacious mental models that do not give an accurate presentation of the environment to fallacious inferences (including confirmation biases, erroneous probabilistic calculations and much more) and errors regarding the range of phenomena towards which the explanations are to be applied.[25]

The explanatory rules within the scientific context emerge and are adopted in an invisible-hand process of innovation and imitation. The point of departure is always a certain body of knowledge, generated

a scientific community maintains critical dialogue. Scientific communities will be objective to the degree that they satisfy four criteria necessary for achieving the transformative dimension of critical discourse: (1) there must be recognized avenues for the criticism of evidence, of methods, and of assumptions and reasoning; (2) there must exist shared standards that critics invoke; (3) the community as a whole must be responsive to such criteria; (4) intellectual authority must be shared equally among qualified practitioners."

Longino (2002, pp. 128ff.) elaborates further these criteria she takes to be necessary to assume the effectiveness of discursive interactions. Although what she calls "critical contextual criticism" is an interesting and important approach, it is untenable in the end, because it refers to an idealized epistemic community, a position very similar, if not identical, with the one proposed by Habermas, whose serious weaknesses are convincingly shown by Herbert Keuth (1993). Her main claim is that "the social position or economic power of an individual or group in a community ought not determine who or what perspectives are taken seriously in that community. Where consensus exists, it must be the result not just of the exercise of political or economic power, or of the exclusion of dissecting perspectives, but a result of critical dialogue in which all relevant perspectives are represented" (2002, p. 131). This requirement, though aiming at a noble aim, is unfortunately simply inapplicable not only to the scientific world but to any world inhabited by real human beings.

25 As McCauley correctly observes, the main difference between scientists and children is that the institutions of science permit the first, but not the second to circumvent their inherent cognitive limitations (2000, p. 67): "Scientists can get around some of their cognitive limitations by exploiting a vast array of tools (such as literacy and mathematical description) and cultural arrangements (such as journals, professional associations, and the division of labor). Children by contrast, mostly work in comparative isolation unaided by these tools, unable to take advantage of such arrangements, and unacquainted with the enormous bodies of knowledge to which scientists have access".

[...] The institution of science does an even better job than either individual scientists or local research teams of getting around cognitive limitations because it is the collective product of an international community of inquirers for whom prestige, fame, and wealth, turn, in no small part, on their seizing opportunities to criticize and correct each other's work. Such communal features of the scientific enterprise establish and sustain norms that govern scientific practice. They also ensure that the *collective* outcome of the efforts and interactions of mistake-prone individuals and small research groups with one another in the long run is more reliable than any of their individual efforts are in the short run."

according to a set of metaphysical presuppositions and in a given institutional order. Ordinary, commonsensical knowledge is always existent as is a given institutional matrix. "We always start from here." In other words, there always exists a body of available knowledge and an institutional a priori, the scheme of existing things. (History of science shows us how the status quo has been shaped and, thus, the constraints which any move beyond the status quo must necessarily respect.) An innovator conceives of a new explanatory problem and attempts to solve it by using what seem to him to be appropriate means of representation, appropriate inferential strategies and the appropriate scope of their application. A novel, successful solution can be attained due to the appropriateness of the representation or the inferential strategies or the scope or to a combination of them. The environmental feedback that the explainer receives comes from the natural environment and the social environment, that is, from the relatively tight institutional structure of science. The "evidence" is a fabricated entity, produced as the outcome of the interaction between a part of nature and a part of social environment. This is plausibly the case in experiments, but also in the simplest cases in which observations are made without the aid of any kind of technological devices – there is always a theoretical component to observations, structuring the sensory input.

The criticism that comes from the social environment of science and is offered to the individual problem-solver can affect the metaphysical presuppositions, the background knowledge, but also, more directly, the quality of the mental representations, the inferences and the chosen scope of application of the explanation. Depending on the quality of the environmental feedback, the problem-solver of the new explanatory problem will revise, reject or simply ignore it. At some point he will choose to adopt the solution that seems to him to be the most appropriate. By reapplying this solution every time the same or a similar explanatory problem emerges, it will become routinized; consequently the problem-solver will adopt a relatively standardized way of providing the particular solution to the particular kind of problem.

In a process of communication and persuasion this standardized solution will sometimes become shared by other members of the scientific community and so the respective rules will be disseminated to a larger group of scientists. It is important to stress that the adoption will sometimes concern only aspects of the standardized solution, that is, the rules of representation or the rules of inference or the rules of scope and not necessarily all of them.[26] Criticism will tend to be more severe, the more the new explanatory problem solutions challenge the intuitions of contemporary scientists (and those that are enforcing the institutional rules prevailing at the time) and the more dissimilar they are from the solutions currently employed. In a process of communication and persuasion some (and sometimes all) rules will be adopted by other group members who become participants to the respective explanatory game. The repetitive use of the rules of representation, rules of inference and rules of scope will turn them into standardized problem solutions to explanatory problems; and they will come to be followed by the game participants in an unconscious way. For those who have accepted and learnt the respective rules, a consensus will have been

---

[26]   The use of telescope, for example, as a legitimate means of representation of the planets had been adopted by astronomers at the time of Galileo even though the inferences that he drew on the basis of his astronomical observations were not shared by many of his contemporaries. See the detailed discussion provided by Feyerabend (1975/2010, p. 82) on the difficulties that Galileo's contemporaries had in admitting the legitimacy of the first telescopic observations of the sky, which were "indistinct, indeterminate, contradictory and in conflict with what everyone can see with his unaided eyes". As Feyerabend provocatively argues, much propagandistic effort was needed in order for Galileo's explanations to be successfully communicated (1975/ 2010, p. 117): "The similarity with the arts which has often been asserted arises at exactly this point. Once it has been realized that a close empirical fit is no virtue and that it must be relaxed in times of change, then style, elegance of expression, simplicity of presentation, tension of plot and narrative, and seductiveness of content become important features of our knowledge. They give life to what is said and help us to overcome the resistance of the observational material. They *create* and maintain interest in a theory that has been partly removed from the observational plane and would be inferior to its rivals when judged by the customary standards. It is in this context that much of Galileo's work should be seen. This work has often been likened to *propaganda* – and propaganda it certainly is. But propaganda of this kind is not a marginal affair that surrounds allegedly more substantial means of defence, and that should perhaps be avoided by the 'professionally honest scientist'. In the circumstances we are considering now, *propaganda is of the essence*."

built about which kinds of representations are to be used and which are not, which rules of inference are permissible and which are not and what the appropriate scope of the explanations is.

If the issue is approached from the view that I have offered here, some traditional apories in the philosophy of science simply evaporate. Thomas Kuhn, in *The Structure of Scientific Revolutions*, seemed to struggle between acknowledging the rules according to which scientists were playing the game and which he took to be explicit, and paradigms which seem to comprise understandings, skills, etc., which he took to be of a primarily tacit character.[27] This has provoked waves of criticism from many quarters focusing – rightly, I think – on the vagueness and polysemy of the concept of paradigm. According to the view proposed here, however, the explanatory games are played by participants who have adopted explicit rules as well as implicit rules, which can be primarily and most effectively shared by means of direct imitation of the practices of others. So, Kuhn is certainly right when he emphasizes the importance of the acquisition of practical skills during the laborious process of scientific education,[28] but there need not be anything mysterious about the learning of the respective problem solutions – this is just the case of the acquisition of primarily procedural knowledge.[29]

[27]   See especially his discussion in Chapter 4.

[28]   See Kuhn (1962/1970, p. 46f.): "Scientists, it should already be clear, never learn concepts, laws, and theories in the abstract and by themselves. Instead, these intellectual tools are from the start encountered in a historically and pedagogically prior unit that displays them with and through their applications. A new theory is always announced together with applications to some concrete range of natural phenomena; without them it would not be even a candidate for acceptance. After it has been accepted, those same applications or others accompany the theory into the textbooks from which the future practitioner will learn his trade. They are not there merely as embroidery or even as documentation. On the contrary, the process of learning a theory depends upon the study of applications, including practise problem-solving both with a pencil and paper and with instruments in the laboratory. If, for example, the student of Newtonian dynamics ever discovers the meaning of terms like 'force', 'mass', 'space', and 'time', he does so less from the incomplete though sometimes helpful definitions in his text than by observing and participating in the application of these concepts to problem-solution." See also Woody (2004) who focuses specifically on the role of representational practices for scientific explanations.

[29]   See Sun et al. (2001, p. 207) for a review of the evidence from cognitive science for the acquisition of rules by agents without being able to verbalize them and Wallis (2008, p. 138) for the evidence from neuroscience.

One important consequence of the dynamics of the explaining activities as depicted within the problem-solving framework outlined in Chapter 8 is exactly the ever-existent routinization of the problem solutions. In the context of interest here, the frequent use of the same rules of representation, inference and scope by the participants in a specific explanatory game leads to the standardization of the offered problem solutions and so the respective rules become largely unconscious.[30] This has two effects. On the one hand, the unconscious use of rules by the explainers (in the context of a cognitive division of labour) serves an unburdening function: the limited cognitive resources of the scientists can be devoted to whatever their current task is and thus can leave the necessary space for creativity to unfold. On the other hand, this collective routinization attenuates the critical potential of the explainers, which, if further cemented by a prevailing dogmatic institutional environment, can inhibit the trial of novel problem solutions: old problem solutions might prevail, even in the light of novel ones that are on offer and might be more appropriate.[31]

---

[30]   It is important to note here that I am not claiming any kind of a priori status for any of these rules. Michael Friedman in his ingenious attempt to develop the *Dynamics of Reason* (2001) attributes to some principles, "basic principles of geometry and mechanics, for example, that define the fundamental spatio-temporal framework within which alone the rigorous formulation and empirical testing of first or base level principles is then possible", the status of a priori. "These relativized a priori principles constitute what Kuhn calls paradigms: at least relatively stable sets of rules of the game, as it were, that define or make possible the problem solving activities of normal science [...]" (2001, p. 45). Friedman (2001, 2002) also opts for the Habermasian conception of communicative rationality (which Habermas himself does not regard as applicable for science, where the instrumental rationality is supposedly dominant). This weakens Friedman's otherwise interesting approach considerably, since the conception of Habermas is very problematic (see the criticism of Keuth (1993) already previously referred to). His more recent extension of the dynamics of reason to include not merely the domain of purely intellectual history, but also the wider cultural context is more promising (Friedman 2011, 2012), but it remains programmatic, despite its ambitious character expressed in the following quote (2012, p. 52): "By integrating some of the highpoints in the development of moral and political philosophy after Kant with the extended narrative in history and philosophy of science I am now in the process of developing, I thereby hope, in the end, to make a transformed – and fundamentally historical – version of Kantian philosophy more directly relevant to the philosophical, scientific, technological, political and spiritual predicaments of our own time."

[31]   Our case study drawn from medicine (Chapter 6, Section 6.3) has been a good exemplification of the theoretical argument that I am proposing here. The more specific

However, notwithstanding the cognitive and institutional mechanics[32] that might lead to a cementation of explanatory rules, it follows from what I have been arguing for throughout the book that there is always a *competition* between alternative explanations offered by scientists. They certainly stand in competition with the different rules used in the different scientific explanatory games, and they stand in a *broader competition* with those offered in religion and in everyday life. There is always a plurality of alternatives to choose from, and this is important to note. The *Myth of the Framework* is indeed a myth, a myth that exaggerates a difficulty into an

point about routinization comes, I think, splendidly into fore in the following quote of William Harvey (ch. VIII): "What remains to be said upon the quantity and source of the blood which thus passes is of a character so novel and unheard of that I not only fear injury to myself from the envy of a few but I tremble lest I have mankind at large for my enemies, so much doth wont and custom become a second nature."

[32] Paul Feyerabend was one of the first to have stressed the role that broader institutional constraints play in the scientific activities of the scientists. See his analysis of the case of Galileo (Feyerabend 1975/2010, p. 119): "There is a further element in this tapestry of moves, influences, beliefs which is rather interesting and which received attention only recently – the role of patronage. Today most researchers gain a reputation, a salary and a pension by being associated with a university and/or a research laboratory. This involves certain conditions such as an ability to work in teams, a willingness to subordinate one's ideas to those of a team leader, a harmony between one's ways of doing science and those of the rest of the profession, a certain style, a way of presenting the evidence – and so on. Not everyone fits conditions such as these; able people remain unemployed because they fail to satisfy some of them. Conversely the reputation of a university or a research laboratory rises with the reputation of its members. In Galileo's time patronage played a similar role. There were certain ways of gaining a patron and of keeping him. The patron in turn rose in estimation only if he succeeded to attract and to keep individuals of outstanding achievement. According to Westfall, the Church permitted the publication of Galileo's *Dialogue* in the full knowledge of the controversial matters contained in it '[n]ot least because a Pope [Urban VIII] who gloried in his reputation as a Maecenas, was unwilling to place it in a jeopardy by saying no to the light of his times', and Galileo fell because he violated his side of the rules of patronage.

Considering all these elements, the 'Rise of the Copernican World-View' becomes a complicated matter indeed. Accepted methodological rules are put aside because of social requirements (patrons need to be persuaded by means more effective than argument), instruments are used to redefine experience instead of being tested by it, local results are extrapolated into space despite reasons to the contrary, analogies abound – and yet all this turns out, in retrospect, to have been the correct way of circumventing the restrictions implied by the human condition. This is the material that should be used to get better insight into the complex process of knowledge acquisition and improvement."

impossibility (Popper 1970, p. 56f.). The decision to accept one set of alternative rules is always a decision to reject another set of rules, and the judgement leading to that decision involves the comparison of the alternatives with nature and with each other.[33] The *criticism* that is possible in scientific explanatory games is much more extended because of the prevalence of the *liberal normative dimension* in which explainers in science usually find themselves operating, which is due to the institutional framework in place.

This fact is concomitant with the fact that there are always normative resources available *outside* the respective explanatory game, both at *different levels* and of *different quality*, normative resources that can be used for purposes of the *normative evaluation* of the rules of the game. As I have argued at the beginning of the book, besides the descriptive task, to which the last four chapters have been devoted, a philosophical theory of explanation also has the normative task to inquire into the grounds that will serve for the evaluation of the different rules of the explanatory games and for adjudicating between different explanations. It is this task that I will turn to in the next two chapters.

---

[33]   This statement considerably corrects, I think, Kuhn' s statement (1962/1970, p. 77): "The decision to reject one paradigm is always simultaneously the decision to accept another, and the judgment leading to that decision involves the comparison of both paradigms with nature *and* with each other."

10 Normative Appraisal: A Procedural Conception

Besides the descriptive task, a philosophical theory of explanation should also tackle the normative task, providing guidance with respect to the evaluation of different explanatory activities and their outcomes. The normative discourse about explanations has been centred around the goodness of the outcomes rather than the goodness of the rules. The alternative account that I am proposing in this and the following chapter focuses on the evaluation of rules and suggests a procedural, multidimensional view.

10.1 THE PROCEDURAL VIEW

The traditional philosophical account of explanation adopts a static stance when trying to develop normative standards to judge the outcomes of explanatory activity. Philosophers of science focus their attention on developing unitary models of what "the successful explanation" is supposed to accomplish. The development of the notion of an "ideal explanatory text" (Railton 1981, pp. 240ff.), for example, is a characteristic attempt to offer an eternal standard for judging the quality of explanations (even if meant to serve only as a regulative ideal). The provision of a general model able to serve as a reference which applies to all times and under all conditions suggests itself if one focuses on explanations as outcomes.

However, the opposing procedural conception is based squarely in the rejection of any claim that the explanations that are good are "out there" waiting to be discovered by scientists or anybody else. The rules of the explanatory game are quite literally made up or created in some participatory process of discussion, criticism and persuasion. Good explanations are those that have emerged because good rules have been followed, and working out the criteria of their goodness is not a once-for-all matter, but rather a continuous enterprise taking place with the participation of different kinds of experts or other participants in the explanatory games and philosophers alike. The elucidation of this participatory process is the fundamentally *naturalistic* element of the approach.

The morale is that it is not possible to provide a general model able to serve as a reference for a successful explanation, which applies at all times and under all conditions. Rather, the concrete specification of the prescriptive rules that govern the explanatory activity must proceed from a critical analysis of the prevailing situation, and it must get by without an atemporal abstract ideal. In other words, the explanatory process is channelled by rules which are themselves changing. This refers to all kinds of rules, including the rules of logic (which are especially important in the context of scientific explanations) since logic, like every other discipline, develops over time.[1] Let us elaborate this general position further.

10.2 THE HIERARCHY OF PROBLEMS, RULES AND VALUES

I would like to attempt a systematization of the discussion by proposing a hierarchical model with *problems* at the *lowest* level, *rules* at the *intermediate* level and *values* at the *highest* level.

I have conceptualized explanatory activity as a special kind of problem solving, in which the goal state is a target to be explained. Problems are very different from "facts" (even "theory-laden facts"), and the solution of a problem cannot be equated with accounting for a fact. This does not preclude the possibility of a great range of problems being facts or states of affairs, but the class of problems includes much more than states of affairs. As Laudan (1977, p. 16) correctly observes, "[a] problem need not accurately describe a real state of affairs to be a problem: all that is required is that it be *thought to be* an actual state of affairs by some

---

[1]   The refinement of the logical aspects of scientific explanation with the aid of first-order predicate logic is, thus, naturally, only an instance in a long evolutionary process rather than the final word on the matter.

agent".[2] In other words, many problems can be literally *counter-factual*: the problem whether angels are male or female is of this type, as is the problem of describing the exact properties of a socialist utopia. Many facts about the world do not present themselves as problems simply because they are unknown. "[A] fact becomes a problem when it is treated and recognized as such; facts, on the other hand, are facts, whether they are ever recognized. The only kind of facts which can possibly count as problems are *known* facts" (Laudan 1977, p. 16f.). Finally, problems can be transformed with the passage of time or they can even cease to be regarded as problems, whereas facts can never undergo a transformation (Laudan 1977, p. 16f.).

The very distinction between problems and facts or, to put it differently, the question of what constitutes a problem (and what constitutes a fact) is very often a contestable issue which is solved by appeal to certain normative rules – these belong to the constitutive rules of an explanatory game, the ones that determine the metaphysical presuppositions, as we have seen in Chapter 6, Section 6.1. In other words, the very fundamental issue of what should count as a problem which must become the target of an explanatory activity is solved by moving one level up in the hierarchy, that is, by appeal to the rules that determine the metaphysical presuppositions. This is the case also with all kinds of divergences of opinion about how a specific explanatory problem is to be solved. Rules of representation, rules of inference

---

[2]   The quotation given in the text continues as follows (Laudan 1977, p. 16): "For instance, early members of the Royal Society of London convinced by mariners' tales of the existence of sea serpents, regarded the properties and behavior of such serpents as an empirical problem to be solved. Medieval natural philosophers such as Oresme, took it to be the case that hot goat's blood could split diamonds and developed theories to explain this counterfactual empirical 'occurrence'. Similarly, early nineteenth century biologists, convinced of the existence of spontaneous generation, took it to be an empirical problem to show how meat left in the sun could transmute into maggots or how stomach juices could turn into tapeworms. For centuries, medical theory sought to explain the 'fact' that bloodletting cured certain diseases. If factuality were a necessary condition for something to count as an empirical problem, then such situations could not count as problems. So long as we insist that theories are designed only to explain 'facts' (i.e., true statements about the world), we shall find ourselves unable to explain most of the theoretical activity which has taken place in science."

and rules of scope are appealed to whenever explanatory problems are at issue. As we have seen, often this appeal is not conscious – to the degree that the different kinds of rules have been adopted after a learning process by the explainers. And often this appeal is not explicit – to the degree that the explainers share a community consensus with respect to which rules are to be used to solve the explanatory problems at hand. The level of rules constitutes the intermediate level in the hierarchy of appraisal.

The virtues and vices of the rules of an explanatory game can themselves be analysed, debated and criticized. This criticism takes place with respect to specific criteria of appraisal encapsulating diverse values. These can be epistemic values, such as accuracy, simplicity, consistency, fruitfulness, etc., or non-epistemic values (i.e., aesthetic, moral, political, social and religious values) such as beauty, honesty, integrity, freedom of expression, piety, etc. There is, naturally, value-pluralism. At this highest level, the critical appraisal of the rules of the explanatory games takes place with reference to these diverse values by participants and non-participants in the game alike. The means of the critical appraisal can, of course, differ. Some explanatory games are characterized by a great dogmatism and the critical appraisal of the rules takes place by violent means – this is often the case with religious explanatory games. The means of critical appraisal used for judging the quality of the rules of modern scientific explanatory games are non-violent – this has become so common that it is regarded as self-evident.

In sum, *explanatory problems* lie at the *lowest* level and are solved with the help of *rules* at the *intermediate* level. Whenever disagreement about the appropriateness of the rules emerges, either about fundamental metaphysical issues (most prominently about what constitutes a problem and which set of problems constitute facts) or about issues of representation or of inference or of explanatory scope, the rules cannot be treated as an unproblematic instrument for conflict resolution. Such controversies are to be resolved by moving one step up the hierarchy, that is, by appeal to epistemic or non-epistemic

Values
(epistemic and non-epistemic)

Explanatory rules
(constitutive, of representation, of inference, of scope)

Explanatory problems

FIGURE 11 The Hierarchical Model

values. The *axiological level of values* is the *highest* level in the hierarchy. Figure 11 summarizes this hierarchical model.

10.3   THE MULTIDIMENSIONALITY OF THE CRITICAL APPRAISAL

This hierarchical model is very appealing as the first step of tackling the normative task, but it must be modified and completed in a second step to provide a more thorough and fine-grained account. The completion concerns both the level of rules and the level of values, but also, most importantly, the way that they are connected.

At the level of rules, *first*, it has been the main tenet throughout my discussion that there is a plurality of rules whose invention and subsequent routinization and adoption lead to the provision of explanations; we have systematized them as (1) constitutive rules, (2) rules of representation, (3) rules of inference and (4) rules of scope. It should be clear, thus, that this systematization enables a fine-grained approach at the level of explanatory rules, and this should be an obvious point by now, not in need of further argument or clarification. We have described in some detail in Chapter 9, Section 9.4. how the rules of the games that children are playing, but also of the religious and scientific games, emerge and are adopted.

Second, a fine-grained approach at the level of values is also an important element of the general conception that I propose. A broad systematization into epistemic and non-epistemic values is important here. Values provide the fixations of normative resources that can help judge the goodness of explanatory rules that have emerged in the context of the historical development of different explanatory games. Since explanatory games are always embedded in broader normative frameworks, as we have seen, it is only on the grounds of both epistemic and non-epistemic values that a more thorough evaluation can take place. Values most importantly provide the normative resources for *judgements across explanatory games* that are permanently unfolding in parallel. A catalogue of epistemic values is contained in Kuhn's classic essay "Objectivity, Value Judgment, and Theory of Choice" (1977), which includes (a) accuracy, (b) consistency, (c) scope, (d) simplicity and (e) fruitfulness.[3] These values have been discussed thoroughly in the literature in the light of the debates on the appropriate criteria for the evaluation of the adequacy of *theories*. (In other words, reflecting the dominance of the view that the centre of explanatory, and indeed scientific, activity lies in the generation and evaluation of sets of statements – usually in the forms of theories or models – which constitute the fundamental if not the exclusive representational resources, most of the debate has concerned the

---

[3] See Kuhn (1977, p. 321f.): "What, I ask to begin with, are the characteristics of a good scientific theory? Among a number of quite usual answers I select five, not because they are exhaustive, but because they are individually important and collectively sufficiently varied to indicate what is at stake. First, a theory should be accurate: within its domain, that is, consequences deducible from a theory should be in demonstrated agreement with the results of existing experiments and observations. Second, a theory should be consistent, not only internally or with itself, but also with other currently accepted theories applicable to related aspects of nature. Third, it should have broad scope: in particular, a theory's consequences should extend far beyond the particular observations, laws, or subtheories it was initially designed to explain. Fourth, and closely related, it should be simple, bringing order to phenomena that in its absence would be individually isolated and, as a set, confused. Fifth – somewhat less standard item, but one of special importance to actual scientific decisions – a theory should be fruitful of new research findings: it should, that is, disclose new phenomena or previously unnoted relationships among those already known. These five characteristics – accuracy, consistency, scope, simplicity, and fruitfulness – are all standard criteria for evaluating the adequacy of a theory. If they had not been, I would have devoted far more space to them in my book, for I agree entirely with the traditional view that they play a vital role when scientists must choose between an established theory and an upstart competitor. Together with others of much the same sort, they provide *the* shared basis for theory choice." For a discussion and elaboration of these epistemic values see McMullin (2008). Laudan (2004) draws a distinction between epistemic and cognitive values, arguing that truth as the epistemic value is one among a series of other cognitive values, such as generality and scope. For a discussion of possible groupings of epistemic/cognitive values see Douglass (2013).

goodness of these representations. This is not the place to engage in the discussion of issues such as what constitutes a theory, whether one should focus on the dyadic relationship between language-like entities and the world or rather on the use of the theories and models to represent reality for a specific purpose and the like.[4] It is merely important that these epistemic values have been used to evaluate theories or models in a quasi-holistic fashion.) Besides, as already noted, there is value pluralism not only with respect to epistemic values but also with respect to non-epistemic values: beauty, freedom of expression, honesty, integrity, etc.[5]

Third, the plurality of rules and the plurality of values allow for a *multidimensional evaluation*. Different sets of rules can be evaluated with respect to different values – this is the main tenet of a fine-grained approach. *Appraisal is always a comparative matter* – we always form evaluative judgements with respect to a value by comparing different alternatives. This comparative approach must be juxtaposed to the traditional approach according to which explanations are judged positively if they fit to an ideal picture outlined by the respective normative theory of explanation. For example, if a maximal unification of different phenomena has been accomplished by a specific explanation, then this is the one to be preferred (Kitcher 1989). Or, if an explanation has accounted for a high-level phenomenon, say inflation, in terms of the lowest ontological level, say molecules, atoms and the like, then this is the one to be preferred (Strevens 2008). The comparative approach suggests that the explanations produced in the explanatory games should be judged with respect to multiple values along the different dimensions. Since it is the following of rules

---

[4]   On this see the important paper of Ronald Giere (2004).

[5]   The distinction between epistemic and non-epistemic values has been drawn by Ernan McMullin in his 1982 presidential address to the Philosophy of Science Association (1983/2012, pp. 701ff.). See also the interesting approach to epistemic and non-epistemic values of Carrier (2013). Brown (2013) offers a good overview of the recent discussion of values in science which has mainly been centred around arguments from underdetermination of theory by evidence and arguments with respect to inductive risk. See also the interesting discussion by Bueter (2015).

that have mainly given rise to the respective explanations, those rules should be the object of the comparative evaluation.

This comparison, as all comparisons, is to be directed towards some kind of ideal. However, the ideal is not the preferred image of a perfect explanation allegedly defined with the aid of sufficient and necessary conditions by a philosopher. The ideal is composed, instead, of the series of epistemic and non-epistemic values encapsulating the normative resources at a high level, which function as Kantian "regulative Ideen". Insofar, it is more abstract than the image of a perfect explanation and more suitable for a thorough and multidimensional evaluation.

Explanatory activity is an epistemic activity and its products can be sensibly evaluated according to epistemic values. However, explanatory games are always embedded in an institutional framework that to a great degree regulates the content of the feedback received by the explainers. (We have given a descriptive account of the different kinds of institutions within which the explanatory games played by children, the religious explanatory games and the scientific explanatory games typically unfold.) The formal and informal institutions comprising the institutional framework can be sensibly evaluated according to non-epistemic values. The image that

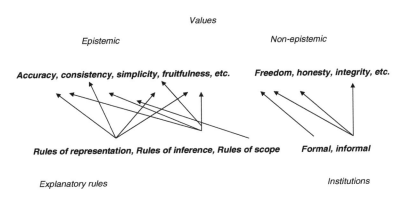

FIGURE 12 The Multidimensional Model

emerges is, thus, one (a) of an evaluation of the different explanatory rules according to different epistemic values and (b) of the different institutions that provide the framework of the explanatory games according to different non-epistemic values. A multidimensional comparative evaluation is the result (figure 12).

11 Explanatory Methodology as Technology

Probably the most important question is whether there is some kind of external criterion or justification process that can be used in order to judge the superiority of the different values that are used in the critical appraisal of the rules of different explanatory games. In other words, is there an ultimate justification for values and other normative structures?

11.1 THE MÜNCHHAUSEN TRILEMMA

The quest for an ultimate justification of values and rules is a manifestation of the vain quest for certainty originating in the idea of a positive, sufficient justification. This general philosophical idea, operative both in epistemology and philosophy of science as well as in ethics and political philosophy, mandates the quest for an adequate foundation, a sufficient justification, for all convictions and principles. The demand for a justification of everything leads to a situation with three alternatives, all of which are unacceptable: the Münchhausen Trilemma. The Baron von Münchhausen was a German nobleman who lived in the eighteenth century and has become famous as a storyteller. In one of his famous stories he managed to pull himself and the horse on which he was sitting out of a swamp by his own hair. Though the main outline of this problem was known in the context of ancient scepticism, more specifically in the context of the discussion of Agrippa's five tropes (Sextus Empiricus, Outlines of Pyrrhonism, Book 1, ch. IE),[1] it was the

---

[1] The five tropes or modes have been among the most famous arguments of ancient scepticism and are attributed to Agrippa by Diogenes Laertius (who lived towards the end of the first century CE). They are also given by Sextus Empiricus in Book 1 of the *Outlines of Pyrrhonism*. See chapter XV – Of the Five Modes (164–169): "The later Sceptics hand down Five Modes leading to suspension, namely these: the first based

German philosopher Hans Albert who first drew attention to how this story by Münchhausen[2] provides a fitting analogy for the following situation to which one is led by a demand for a justification for everything:

One must choose [...] between

> on discrepancy, the second on regress *ad infinitum*, the third on relativitiy, the fourth on hypothesis, the fifth on circular reasoning: That based on discrepancy leads us to find that with regard to the object presented there has arisen both amongst ordinary people and amongst the philosophers an interminable conflict because of which we are unable either to choose a thing or reject it, and so fall back on suspension. The Mode based upon regress *ad infinitum* is that whereby we assert that the thing adduced as a proof of the matter proposed needs a further proof, and this again another, and so on *ad infinitum* so that the consequence is suspension, as we possess no starting-point for our argument. The Mode based upon relativity, as we have already said, is that whereby the object has such or such an appearance in relation to the subject judging and to the concomitant percepts, but as to its real nature we suspend judgment. We have the Mode based on hypothesis when the Dogmatists, being forced to recede *ad infinitum*, take as their starting-point something which they do not establish by argument but claim to assume as granted simply and without demonstration. The Mode of circular reasoning is the form used when the proof itself which ought to establish the matter of inquiry requires confirmation derived from that matter; in this case, being unable to assume either in order to establish the other, we suspend judgment about both."

2   Gottlob Frege in the preface of *Grundgesetze der Arithmetik* (1893/1998, p. XIX) had already referred to Baron von Münchhausen in the context of the quest for an ultimate justification: "Wo ist denn hier der eigentliche Urboden, auf dem Alles ruht? oder ist es wie bei Münchhausen, der sich am eigenen Schopfe aus dem Sumpfe zog?" Karl Popper in the *Logik der Forschung* (1934) has discussed a similar but not identical problematic with reference to Jacob Fries's *Neue oder anthropologische Kritik der Vernunft* (1828–1831), which he baptized as "Fries's Trilemma". Here is the quote from the English translation, i.e., the *Logic of Scientific Discovery* (§25, p. 75): "The problem of the basis of experience has troubled few thinkers so deeply as Fries. He taught that, if the statements of science are not to be accepted *dogmatically*, we must be able to *justify* them. If we demand justification by reasoned argument, in the logical sense, then we are committed to the view that *statements can be justified only by statements*. The demand that all statements are to be logically justified (described by Fries as a 'predilection for proofs') is therefore bound to lead to an *infinite regress*. Now, if we wish to avoid the danger of dogmatism as well as an infinite regress, then it seems as if we could only have recourse to *psychologism*, i.e. the doctrine that statements can be justified not only by statements but also by perceptual experience. Faced with this *trilemma* – dogmatism vs. infinite regress vs. psychologism – Fries, and with him almost all epistemologists who wished to account for our empirical knowledge, opted for psychologism. In sense-experience, he taught, we have 'immediate knowledge': by this immediate knowledge, we may justify our 'mediate knowledge' – knowledge expressed in the symbolism of some language. And this mediate knowledge includes, of course, the statements of science."

1. an *infinite regress*, which seems to arise from the necessity to go further and further back in the search for foundations, and which, since it is in practice impossible, affords no secure basis;
2. a *logical circle* in the deduction, which arises because, in the process of justification, statements are used which were characterized before as in need of foundation, so that they can provide no secure basis; and finally
3. the *breaking-off of the process* at a particular point, which, admittedly, can always be done in principle, but involves an arbitrary suspension of the principle of sufficient justification.

Since both an infinite regress and a circular argument seem clearly unacceptable, one is inclined to accept the third possibility, for the simple reason that no other way out of the situation is thought to be possible. Of statements where one is prepared to break off the foundation process, it is customary to use words such as 'self-evident,' 'self-authenticating,' 'based upon *immediate* knowledge' – upon intuition or experience that is – or in some other way to render palatable the fact that one is prepared to break off the foundation regress at some particular point and suspend the foundation postulate with respect to this point, declaring it to be an Archimedian point (of knowledge). [...] If such a conviction or statement is called a dogma, then our third possibility is revealed as something one would have least expected in a solution to the foundation problem: justification by *recourse to a dogma*.

(Albert 1968/1985, p. 18f)

Typically the traditional philosophical theory of explanation dogmatically suspends the argumentation at some point in favour of the ultimate "given" that the respective philosopher favours. This dogmatism and the Münchhausen Trilemma can be avoided by substituting the *principle of critical examination* for the *principle of sufficient justification*. In accord with this principle, a discussion of the problems of epistemology, philosophy of science, or ethics need not refer back to ultimate reasons in order to be convincing or "rational". Instead, problems that arise in the sphere of cognition or in the sphere of praxis are to be discussed and solved in light of the already existing solutions. Solutions to new types of problems of any sort require creativity and imagination, and they are not worked out in a social and mental vacuum. The application of the principle of critical examination means that solutions are to be creatively discovered, they are to be weighted in reference to certain values and standards, and, on this basis, the preferred solutions to problems are to be decided upon. Solutions are not judged to be good or rational by virtue of being based on certain knowledge or ultimate values. Instead, our solutions in all areas of cognition and action are fallible, but they can be improved by critical discussion.

11.2 BETWEEN DOGMATISM AND RELATIVISM: THE MIDDLE-GROUND POSITION

The idea that certain models of explanation are immanently rational or good is a dogma. In contrast, the critical discussion of the rules of explanatory games aiming at their modification and revision on the basis of a weighting of the various feasible alternatives with regard to different criteria and values is a way of escaping dogmatism, without sliding into any kind of anarchic position of "anything goes". This middle-ground position between dogmatism and relativism consists of three essential ingredients: pluralism, fallibilism and revisionism. Let me specify very briefly these three ingredients.

A pluralistic account acknowledges positively the existence of many values, epistemic and non-epistemic alike. *Pluralism* is fundamentally different than relativism since pluralism does not involve a renunciation of judgement and commitment as relativism does (Chang 2012, p. 261). A pluralist engages with what she disagrees with, poses critical arguments and proposes alternatives – an attitude very far away from the relativistic "whatever". Besides, relativism does not necessarily imply pluralism: the requirement to treat equally all alternatives that exist does not entail the requirement to have many alternatives. If all agree on one alternative and there is no active search for other alternatives, relativism has nothing to oppose to monism (Chang 2012, p. 261).

Fallibilism is the position that all our knowledge, activities, principles, positions and rules are prone to error. In all areas of cognition and action human beings constantly make mistakes, but they are able to learn from them. A fallibilist treats all problem solutions as hypothetical: she provisionally accepts them instead of searching for a final justification for them. The fallibilistic attitude can be applied to all areas of human activity: science, politics and even religion. (The existence of God can be treated as a hypothetical postulate to be critically discussed, for example.) What concerns us more here, criteria and values can all be accepted as hypotheses amenable to criticism. Fallibilism and the acceptance of the principle of critical examination naturally lead to revisionism.

Revisionism is the position that all our beliefs and problem solutions, fallible themselves, are amenable to revision – and possibly to a progressive one. Revisionism is a corollary of the procedural view adopted – any static ideal would not be compatible with it.

According to this middle-ground position explanatory methodology can be viewed as a technological discipline. As is well known, a technology operates with *hypothetical* rather than *categorical imperatives*. Having accepted that the quest for certainty is vain and that the attempts to provide final justifications for our values are futile, the main endeavour consists in provisionally accepting a series of values and normative criteria that have emerged and then inquiring into how the different rules help to achieve them. To judge the performance of different sets of rules of explanatory games, it is only necessary to hypothetically presuppose certain values, often specified with the help of certain performative criteria, and then to investigate the degree to which the explanatory activities guided by these rules can fulfil these criteria. Accuracy is such a value (which can be further specified in the specific context, say as qualitative or quantitative accuracy, empirical fit, etc.), and a critical discussion of certain rules of an explanatory game can take place with respect to whether at a certain point of time they are accurate or close to being accurate. However, if accuracy is not accepted as a criterion or epistemic value, then their quality can be judged according to other criteria, such as, for example, beauty. And it is very important to stress that the critical discussion concerns also the specification of a certain value in a specific context. Accuracy can be specified in the specific context, say as quantitative or qualitative accuracy.[3] Simplicity can be specified as semantic or syntactic simplicity[4] or cognitive ease, etc.

Favouring explanatory methodology as a technological discipline does not force us to ostracize the domain of values to a dogmatic or relativistic heaven. Values are not to be viewed as impervious to critical discussion as the positivistic dogma demanded for decades.[5] Nor are they ostracized to an indifferent relativistic

[3] For an example of the specification of accuracy in a qualitative way in connection with the idea of truth in a context of comparing alternative interpretations of a text – an evidently hard case – see Mantzavinos (2005, pp. 132ff.). With respect to visual representations, as for example medical drawings, accuracy can be specified as the standard of pictorial fidelity. Such a specification would imply that accuracy is distinct from *precision*. A drawing can be perfectly *accurate* in depicting in a black and white system the figures in a scene, despite the fact that it is *imprecise* with respect to colour. Besides, accuracy would not necessarily entail *actuality*. As Greenberg (2013, p. 224) notes: "A picture may accurately depict a merely possible scene just as well as an actual one, as illustrated by the use of architectural drawings in the evaluation of merely proposed building plans."

[4] Sober has proposed to view simplicity of theories as the minimization of something, for example (2002, p. 15): "To strive for simplicity in one's theories means that one aims to minimize something. The minimization might involve a semantic feature of what a set of sentences says or a syntactic feature of the sentences themselves. An example of the first, semantic, understanding of simplicity would be the idea that simpler theories postulate fewer causes, or fewer changes in the characteristics of the objects in a domain of inquiry. An example of the second, syntactic understanding of simplicity would be the idea that simpler theories take fewer symbols to express, or are expressible in terms of a smaller vocabulary."

[5] See the classic exposition in Carnap's "Überwindung der Metaphysik durch logische Analyse der Sprache" (1932, p. 237): "Die logische Analyse spricht somit das Urteil der Sinnlosigkeit über jede vorgebliche Erkenntnis, die über oder hinter die Erfahrung greifen will. Dieses Urteil trifft zunächst jede spekulative Metaphysik, jede vorgebliche Erkenntnis aus *reinem* Denken oder aus *reiner* Intuition, die die Erfahrung entbehren zu können glaubt. […] Weiter gilt das Urteil auch für alle *Wert- oder Normphilosophie*, für jede Ethik oder Ästhetik als normative Disziplin. Denn die objektive Gültigkeit eines Wertes oder einer Norm kann ja (auch nach Auffassung der Wertphilosophen) nicht empirisch verifiziert oder aus empirischen Sätzen deduziert werden; sie kann daher überhaupt nicht (durch einen sinnvollen Satz) ausgesprochen werden. Anders

universe.[6] Values as normative principles of highest generality can be themselves critically debated.[7] However, the outcomes of such discussions are always of a provisional character, the results of human endeavour themselves, amenable to further debate and criticism. An Archimedean point of departure does not exist in this case; nor does it in any other case. There can just be moments or longer periods of consensus about what are supposed to be the highest virtues of the rules of the game, crystallizing the fallible outcomes of the ongoing

gewendet: Entweder man gibt für 'gut' und 'schön' und die übrigen in den Normwissenschaften verwendeten Prädikate empirische Kennzeichen an oder man tut das nicht. Ein Satz mit einem derartigen Prädikat wird im ersten Fall ein empirisches Tatsachenurteil, aber kein Werturteil; im zweiten Fall wird er ein Scheinsatz; einen Satz, der ein Werturteil ausspräche, kann man überhaupt nicht bilden." Hempel used to be less pronounced and allowed for the possibility of making changes in what he called categorical judgements of values, though he did not seem to have acknowledged the possibility of a rational critical discussion of values. See the closing passage of his essay on *Science and Human Values* (1960) in Hempel (1965, p. 96): "Thus, if we are to arrive at a decision concerning a moral issue, we have to accept some unconditional judgments of value; but these need not be regarded as ultimate in the absolute sense of being forever binding for all our decisions, any more than the evidence statements relied on in the test of scientific hypothesis need to be regarded as forever irrevocable. All that is needed in either context are *relative* ultimates, as it were: a set of judgments – moral or descriptive – which are accepted at the time as not in need of further scrutiny. These relative ultimates permit us to keep an open mind in regard to the possibility of making changes in our heretofore unquestioned commitments and beliefs; and surely the experience of the past suggests that if we are to meet the challenge of the present and the future, we will more than ever need undogmatic, critical, and open minds."

6   See the position of the strong programme in the sociology of science, e.g., Barnes *et al.* (1996). I cannot go into details here and offer a discussion of the Science Wars, but see Sokal and Bricmont's attempt to show the absurdities to which relativism that verges on postmodernism can lead in their *Impostures Intellectuelles* (1997).

7   See Laudan (1984, p. 61f.): "More generally, there are plenty of cases of axiological disagreement in which there is ample scope for fully rational individuals to disagree about goals even when they fully agree about shared examples. But that is a far cry from the familiar claim [...] that virtually all cases of disagreement about cognitive values are beyond rational resolution. It is crucial for us to understand that scientists do sometimes change their minds about their most basic cognitive ends, and sometimes they can give compelling arguments outlining the reasons for such changes. In this regard, disagreements about goals are exactly on a par with factual and methodological disputes. Sometimes they can be rationally brought to closure; other times, they cannot. But there is nothing about the nature of cognitive goals which makes them intrinsically immune to criticism and modification."
See also the discussion of trade-offs between different dimensions of the goodness of an explanation by Ylikoski and Kuorikoski (2010).

discussion. There is, to put it differently, no ultimate justification of the rules but a provisional consensus, if at all, about the most general normative standpoints and the highest principles, a consensus amenable to revisions.

11.3 EXPLANATORY METHODOLOGY AS A RATIONAL HEURISTIC

But the technological character of explanatory methodology that I would like to stress here is not dependent on the outcome of the ongoing discussion about values. Provisionally accepting certain values enables a rational discussion of the appropriateness of different rules with respect to attaining the values. It is a debate of the standard genre of a normal means-ends framework: rule X (of type a), compared with rules Y and Z (of the same type) is more accurate; rule A (of type b), compared with rules B and C (of the same type) is simpler, etc. The problematic of a *final justification* has been transformed to the problematic of a *comparative evaluation*[8] of a multidimensional character.

Unambiguous judgements with respect to whether the respective rules are close to the attainment of the different values is possible. The danger of pervasive relativism that seems to follow from Kuhn's stress on the subjective element of such judgements seems to be exaggerated.[9] As Laudan (1984, p. 31f.) correctly observes in this context:

Suppose that a scientist is confronted with a choice between specific versions of Aristotle's physics and Newton's physics.

8 See Albert (1982, p. 10f.).

9 Kuhn in the aforementioned essay "Objectivity, Value Judgment, and Theory of Choice" (1977), acknowledges the role that rules play in the choice of scientific theories, but tends to draw relativistic conclusions (p. 324f.): "When scientists must choose between competing theories, two men fully committed to the same list of criteria for choice may nevertheless reach different conclusions. Perhaps they interpret simplicity differently or have different convictions about the range of fields within which the consistency criterion must be met. Or perhaps they agree about these matters but differ about the relative weights to be accorded to these or to other criteria when several are deployed together. With respect to divergences of this sort, no set of choice criteria yet proposed is of any use. One can explain, as the historian characteristically does, why particular men made particular choices at particular times. But for that purpose one must go beyond the list of shared criteria to characteristics of the individuals who make the choice. One must,

Suppose moreover that the scientist is committed to observational accuracy as a primary value. Even granting with Kuhn that "accuracy" is usually not precisely defined, and even though different scientists may interpret accuracy in subtly different ways, I submit that it was incontestable by the late seventeenth century that Newton's theory was empirically more accurate than Aristotle's. Indeed, even Newton's most outspoken critics conceded that his theory was empirically more accurate that all its ancient predecessors. Similarly, if it comes to a choice between Kepler's laws and Newton's planetary astronomy – and it does come to such a choice since the two are formally incompatible – and if our primary standard is, say, scope or generality of application (another of the cognitive values cited by Kuhn), then our preference is once again dictated by our values. At best, Kepler's laws apply only to large planetary masses; Newton's theory applies to all masses whatsoever. Under such circumstances the rule "prefer theories of greater generality" gives unequivocal advice.

Such unambiguous judgements are also possible when one proceeds to comparisons between rules that are employed in different

that is, deal with characteristics which vary from one scientist to another without thereby in the least jeopardizing their adherence to the canons that make science scientific. Though such canons do exist and should be discoverable [...], they are not by themselves sufficient to determine the decisions of individual scientists. For that purpose the shared canons must be fleshed out in ways that differ from one individual to another." In his discussion of Kuhn's position in *Science and Values* (1984), Laudan correctly observes, however (p. 31): "If all that Kuhn is saying here is that general rules and values underdetermine choice [...], one might not quarrel with his claim; but, as his subsequent discussion makes clear, he is asserting [...] that preferences are underdetermined as well. He puts it this way: 'every individual choice between competing theories depends on a mixture of objective and subjective factors, of shared and individual criteria.' Kuhn thinks that this state of affairs must be so because, according to his analysis, the 'objective criteria shared by scientists cannot justify one preference to the exclusion of another'. Since, as Kuhn knows perfectly well, scientists do in fact voice theory preferences, he takes this as evidence that they must be working with various individual and idiosyncratic criteria that go well beyond the shared ones. Without the latter, he seems to say, how could scientists ever have preferences? But neither Kuhn nor anyone else has shown, either in fact or in principle, that such rules and evaluative criteria as are shared among scientists are generally or invariably insufficient to indicate unambiguous grounds for preference of certain theories over others."

explanatory games. Take the example that has been presented in Chapter 7 of how the mythical, religious and scientific explanatory games dealt with cosmogony. Recall that these were the cases of explanatory games that have the same explananda, but have different rules determining what must be taken as given and what the metaphysical presuppositions are. Now, one can proceed to unambiguous judgements about the degree of accuracy of the rules of representation used in the three explanatory games unfolding in parallel: algebraic formulas and computer graphs used in the branch of contemporary physics called cosmology are clearly more accurate than the narratives in Hesiod's Theogony and in the Book of Genesis. This judgement is independent on the outcome of the discussion on metaphysical presuppositions employed in the different games.

It should by now have become clear that explanatory methodology can be construed as a technological discipline: explanatory problems are to be solved by rules which can be evaluated with respect to different values which are accepted only hypothetically.[10] It is

---

[10] Explanatory methodology can thus be constructed as a genre of a rational praxis. See Albert (1978, p. 31): "Die in den verschiedenen Kulturbereichen – Wissenschaft, Technik, Recht oder Kunst – dominierenden Bedürfnisse haben zu unterschiedlichen Normierungen geführt, die nicht auf einen Nenner gebracht werden müssen. *Regulative Ideen* oder Ideale wie die der *Wahrheit*, der *Gerechtigkeit* oder der *Schönheit* sind bereichsspezifisch und nicht unbedingt aufeinander reduzierbar, wenn auch nicht auszuschließen ist, daß sich aus Problemlösungen eines bestimmten Bereichs ergeben. Wer eine Aufgabe der Philosophie darin sieht, eine Brücke zwischen diesen Bereichen zu schlagen, muß die Möglichkeit einer konstruktiven oder kritischen Verwendung der Resultate verschiedener Arten menschlicher Praxis füreinander ins Auge fassen.

"Wir können uns aber nicht mit regulativen Ideen der angegebenen Art begnügen, weil sie [...] nicht ohne weiteres *Kriterien* involvieren müssen, die man unmittelbar zur Bewertung von Problemlösungen heranziehen kann. Es ist also nötig, aus den für gute Lösungen konstitutiven Idealen im Zusammenhang mit unserem sonstigen Wissen *Maßstäbe* für die *komparative Bewertung* tatsächlich vorliegender Lösungen zu entwickeln. Auch das ist jeweils eine bereichsspezifische Aufgabe. Die Anwendung solcher Kriterien auf die vorhandenen Lösungen muß aber keineswegs zu einer eindeutigen *Entscheidung* für eine bestimmte Lösung führen, denn es kann unter Umständen *mehrere Maßstäbe* dieser Art geben, die zu jeweils *verschiedenen Ordnungen* zwischen den betreffenden Alternativen führen. Dann muß eine *Gewichtung* der Kriterien herbeigeführt werden, für die unter Umständen übergeordnete Wertgesichtspunkte geltend gemacht werden können. Man sieht, daß eine rationale Praxis Entscheidungen auf verschiedenen Ebenen voraussetzt, die

possible to solve explanatory problems by following different rules. It is of course also possible to evaluate different kinds of rules with respect to different values. For example, the rules of inference of a certain explanatory game can be evaluated with respect to accuracy, but also with respect to beauty. One can have a set of rules of inference that satisfy aesthetic criteria (say highly complex mathematical calculi), but fail to give an accurate account of the phenomena to be explained. The hypothetical imperatives employed are of the following type: if one wants to attain an accurate explanation of phenomenon A, then one should employ rules of inference that provide exact and precise information about the structure of the explanandum phenomenon.

In yet other terms, more general and abstract, this explanatory methodology can be conceptualized as a *rational heuristic*. The development of classical methodology within the framework of classic rationalism is characterized by the fact that the ideal of certain scientific knowledge was tied to the search for a rational heuristic, an *ars inveniendi*, which was supposed to complement the *ars judicandi*, the rational art of proof and justification (Albert 1982, p. 40). Whereas the *ars judicandi* was designed to establish the validity of knowledge claims, the *ars inveniendi* was supposed to provide the discovery of certain knowledge, but both were developed under the influence of the ideal of certain and secure knowledge. The classic rationalism of the modern era thus required a *heuristic* in the form of an *algorithm*, and sometimes even of a *calculus*.

This distinction between the art of invention and the art of justification, though having undergone some transformation, has been largely retained in the course of the development of the

prinzipiell alle als revidierbar angesehen werden können. Sie setzt außerdem Phantasieleistungen aller Art voraus, denn auch die in Betracht kommenden Wertgesichtspunkte müssen entworfen werden, soweit man sich nicht in dieser Hinsicht auf die Tradition verlässt. Das würde aber nur bedeuten, daß man an frühere Kulturleistungen dieser Art anknüpft.

Die menschliche Problemlösungstätigkeit ist also stets konstruktiv und kritisch *zugleich*. Wer ein Problem lösen will, der zielt darauf ab, eine *Lösung* zu finden, *die der Kritik standhält*."

methodology and philosophy of science. The distinction between the context of discovery and the context of justification was a centrepiece of logical positivism and of the Popperian legacy, and it remains in the arsenal of many contemporary philosophers of science.[11] This development went hand in hand with the exclusion of the heuristic altogether, since a new consensus emerged that no algorithmic form for generating new certain knowledge was possible. Heuristic has been excluded from methodology and has been largely assigned to the *domain of the irrational*.[12]

However, this development has not been proven very fruitful (Nickles 2006). On the one hand, decades of discussion after the heyday of logical positivism have brought to light the impossibility of unequivocally making supposedly rational decisions about the validity of scientific knowledge by applying specific algorithms. Even though there are still some attempts to construct formulas designed to replace controversies with calculations (also in the philosophical theory of explanation[13]), these are useless, in the sense that practically no working scientist pays any attention to them. On the other hand, bracketing the problematic of invention and discovery from methodology and relinquishing it to psychology or, even worse, to

---

[11] Of course, this distinction has been recently repeatedly challenged. See e.g., Arabatzis (2006).

[12] See the remark of Popper in the first chapter of *Logik der Forschung* (1934, p. 7): "Unsere Auffassung [...], daß es eine logische, rational nachkonstruierbare Methode, etwas Neues zu entdecken, nicht gibt, pflegt man oft dadurch auszudrücken, daß man sagt, jede Entdeckung enthalte ein ‚irrationales Moment,' sei eine ‚schöpfersiche Intuition' (im Sinne Bergsons); ähnlich spricht Einstein über ‚...das Aufsuchen jener allgemeinsten...Gesetze, aus denen durch reine Deduktion das Weltbild zu gewinnen ist. Zu diesen...Gesetzen führt kein logischer Weg, sondern nur die auf Einfühlung in die Erfahrung sich stützende Intuition."

[13] See e.g., Schupbach and Sprenger (2011) and Crupi and Tentori (2012). In a devastating criticism of such attempts Glymour (2015, p. 603) concludes: "The various formulas for explanatory virtues do not capture much of scientific practice or accord with judgments of the explanatory power of historical scientific cases from a present point of view; they do not accord with common judgments of explanatory power of causes for effects; they do not describe psychological judgment of explanatory power; they do not provide hypothesis tests or useful hypothesis selection criteria. They are not, in a phrase, scientifically serious, and philosophy of science that is not scientifically serious is not serious philosophy."

irrationalism, seems also to be a relic of the heyday of logical positivism, and one which poses an obstacle to a productive methodology of science.

Basing explanatory, and more broadly, scientific methodology *merely* on the interplay of logic and evidence is thus insufficient, since it unduly brackets the *role of imagination* in the cognitive praxis[14] and probably most importantly the *role of choice*. Decisions, that is, human decisions by real-world human beings, are permanently involved in many stages of the process of the comparative evaluation of alternatives that are created with the aid of imagination. Dethroning Olympian rationality as an impossible ideal does not necessarily lead to irrationalism or relativism. (Explanatory) methodology can be conceptualized as a rational heuristic that offers guidance to the cognitive praxis which is oriented towards the discovery of (explanatory) problem solutions.[15]

---

[14] See Albert (1987, p. 83): "Daraus ergibt sich aber gleichzeitig die große *Bedeutung der Phantasie* für eine solche Erkenntnispraxis. Die Entwicklung alternativer Erklärungsansätze und sogar die Entdeckung von Gegenbeispielen ist nämlich vor allem auch eine Sache methodisch disziplinierter Phantasie, und zwar theoretischer Phantasie, die auf das Ziel adäquater Erklärung, und experimenteller Phantasie, die auf das Ziel gerichtet ist, adäquate Prüfungssituationen aufzufinden. Da die Suche nach guten Erklärungen und erklärungskräftigen Theorien prinzipiell unabgeschlossen ist, kann es kein System von Regeln geben, das endgültige Entscheidungen ermöglicht. Im Rahmen eines konsequenten Fallibilismus ist nur eine Methodologie vertretbar, die ausschließlich revidierbare Entscheidungen kennt. Aus dem oben Gesagten geht hervor, daß die Charakterisierung der wissenschaftlichen Methode durch den Hinweis auf ein Zusammenspiel von *Logik* und *Erfahrung* nicht angemessen ist, da dabei die Rolle der Phantasie nicht berücksichtigt ist."

[15] As Hans Albert correctly points out (1987, p. 87f.): "Die Idee, man müsse deshalb, weil man scharf zwischen Fragen der Geltung und der Genese zu unterscheiden habe, das Entdeckungsproblem aus der Methodologie ausscheiden, hat meines Erachtens zu einer Fehlorientierung der Methodendiskussion geführt. Es geht hier ja gar nicht um die Analyse der Genese von Erkenntnissen – also um die Untersuchung bestimmter Tatsachen des Erkenntnisgeschehens –, sondern um die *Anleitung einer Erkenntnispraxis*, die *auf Entdeckungen aus* ist, und es geht auch nicht um die Konstatierung der Geltung irgendwelcher Resultate dieses Geschehens, sondern um die *Anleitung einer Erkenntnispraxis*, die *nicht beliebige Phantasieprodukte als Entdeckungen gelten lassen* kann und daher entsprechende *Bewertungen* und *Entscheidungen* vornehmen muß. Eine Methodologie, die dazu geeignet sein soll, muß also Gesichtspunkte entwickeln, die eine solche Orientierung der Erkenntnispraxis ermöglichen. Sie ist *nicht deskriptiv* – etwa im Sinne einer Beschreibung der Sitten und Gebräuche im Bereich der wissenschaftlichen

The rationality of the proposed explanatory methodology does not lie in the proof or defence of a specific algorithm or justificatory procedure as the most adequate one to solve the problems at hand.[16] The rationality lies instead in laying the ground for the *construction* of alternatives in the form of specifying and clarifying the working properties of the different kinds of explanatory rules and for their *criticism* in the form of highlighting the specific characteristics of a multidimensional evaluation of the different rules as they are embedded in the broader institutional framework. This is a procedural rationality which does not prescribe the content of the decisions which are ultimately taken by the participants of the explanatory games themselves, but highlights instead the *individual* and *collective* conditions that must prevail in order for these decisions and judgements to be informed and adequate with respect to the goals and values that they are aiming at.

A rational heuristic of this kind does not reduce explanatory methodology to calculations that do not leave any space for imagination and genuine decision making.[17] The recent move by some

Forschung –, sie ist auch *nicht normativ*, wie die meisten Theoretiker annehmen, die den Deskriptivismus ablehnen. Sie hat vielmehr *technologischen* Charakter, weil sie methodische Gesichtspunkte enthält, die auf die Erzielung bestimmter Erkenntnisleistungen abgestellt sind. Sie kann daher, *soweit* diese Ziele akzeptiert sind, zur *Normierung* der Erkenntnispraxis beitragen. Daher kann man sie wohl am besten als eine *rationale Heuristik* auffassen, die dazu da ist, diese Praxis in Richtung auf den *Erkenntnisfortschritt* anzuleiten, sie im Sinne dieser Zielsetzung zu normieren. Da diese Zielsetzung keineswegs sakrosankt ist, obwohl sie den tatsächlichen Gang der Wissenschaft – zumindest in der Neuzeit – weitgehend entspricht, ist auch sie natürlich der Diskussion zugänglich."

[16] See Albert (1987, p. 89): "Die Pointe eines methodischen Kritizismus, der an die Stelle der klassischen Methodologie treten kann, liegt nicht darin, jeweils eine bestimmte Problemlösung durch ein Begründungsverfahren irgendwelcher Art endgültig auszuzeichnen, sondern darin, *alternative* Lösungsvorschläge – Theorien, Erklärungen, Beschreibungen – zu *konstruieren* und zu *kritisieren*, das heißt sie in bezug auf ihre komparative Leistung zu beurteilen, um eine *Entscheidung* zwischen ihnen zu ermöglichen."

[17] This was already correctly seen by Max Weber in his classic essay of 1919, *Wissenschaft als Beruf:* "Nun ist es aber Tatsache: daß mit noch so viel von solcher Leidenschaft, so echt und tief sie auch sein mag, das Resultat sich noch lange nicht erzwingen läßt. Freilich ist sie eine Vorbedingung des Entscheidenden: der >>Eingebung<<. Es ist ja wohl heute in den Kreisen der Jugend die Vorstellung sehr verbreitet, die

philosophers of science to present formal models of decision making imported by neoclassical economics and formal decision theory simulates an inexistent precision in conditions of pervasive uncertainty.[18] These conditions, which are the prevailing conditions whenever a choice of rules takes place, are characterized by *genuine uncertainty* rather than *parametric uncertainty*, where probabilities can be calculated and assigned to options. Choice cannot remain choice and its outcomes be predetermined.[19] Choice requires the imaginative faculties of the mind and involves a creative element that must be acknowledged and must find its place in any methodology.

Of course, the outcomes of those decisions are themselves fallible. Within the premises of the problem-solving model presented in Chapter 8 choices take place when new problems are to be dealt with. As with all problem-solving activities and even more so in the case of choice where constructive imagination is involved, errors are possible and common. Institutionalizing the possibility of criticism is thus the best means to facilitate the correction of errors when choices of rules are involved. Anchoring of the freedom of criticism in the institutional framework of the society is the *collective* condition that must prevail and enables procedural rationality to manifest itself.

> Wissenschaft sei ein Rechenexempel geworden, das in Laboratorien oder statistischen Kartotheken mit dem kühlen Verstand allein und nicht mit der ganzen >>Seele<< fabriziert werde, so wie >>in einer Fabrik<<. Wobei vor allem zu bemerken ist: daß dabei meist weder über das, was in einer Fabrik, noch was in einem Laboratorium vorgeht, irgendwelche Klarheit besteht. Hier wie dort muß dem Menschen etwas – und zwar das Richtige – *einfallen*, damit er irgend etwas Wertvolles leistet. Dieser Einfall aber lässt sich nicht erzwingen. Mit irgendwelchem kalten Rechnen hat er nichts zu tun" (Weber 1985, p. 589).

18   See Kitcher's fitting remark (2011b, p. 36): "There is a strong temptation to believe that to show a decision is aligned with canons of good reason and good judgment requires producing explicit canons and demonstrating how the decision exemplifies them. To resist the temptation, we should recall the many complex instances in which good judges are unable to articulate precise rules that guide them – detailed immersion in complex legal decisions can bring conviction that the judgment was a good one, even though there is no body of explicit theory to which one can appeal to support the conviction. We should think of the resolution of complex scientific decisions in a similar fashion." I have dealt with the case of judicial decisionmaking in Mantzavinos (2007).

19   See my detailed discussion of the choice aspect within the general problem-solving framework in Mantzavinos (2001, ch. 4).

All we avail ourselves of as participants to the critical discussion is the possibility of framing informed judgements about these issues and defending them with arguments. A holy grail to help us find the Scientific Method does not exist. Our fallible judgements are all that we have, and at some point we have to make choices about the rules we prefer – not necessarily rejecting the rest (there can be also a second and a third prize, not only a first prize). And it is important to stress again that there is no algorithm relieving us from the necessity to make the choices. Controversies cannot be substituted by calculations – we still have to form judgements and decide which calculations to adopt.

11.4   EXPLANATORY PROGRESS AND EXPLANATORY REGRESS

"Is there any progress?"
"Progress with respect to what?"

A source of permanent confusion regarding questions of progress in general and scientific progress in particular stems from the lack of acknowledgment of the simple issue that judgements about progress are parasitic on the specification of goals – progress makes sense only with respect to the satisfaction or attainment of a specific goal or aim. In the case of interest here, one can sensibly speak about progress with respect to the attainment of a specific value. Does a sequence of refinements, generalizations and extensions of the rules of representation $R_1, R_2 \ldots R_n$ used in the explanatory game A move the explainers closer to achieving the goal of empirical fit than they were before? Then progress relative to that goal state has occurred – if not, then not. Does a sequence of refinements, generalizations and extensions of the rules of inference $I_1, I_2 \ldots I_n$ used in the explanatory game A move the explainers closer to achieving the goal of consistency? Then progress relative to that goal state has occurred – if not, then not. And so on.[20]

---

[20]   Laudan (1984, p. 65), arguing a similar point, correctly points out that "[w]riters on the idea of progress (e.g. Kuhn) have failed to see this point because they seem to assume that progress must always be judged relative to the goals of the agents who performed

It should be obvious that the same set of a specific kind of rules can be progressive with respect to one value and regressive with respect to another[21] – for example the development of a set of rules of representation might be progressive with respect to beauty but regressive with respect to empirical fit. And from the different kinds of rules used in the same explanatory game some can be progressive with respect to the same value and some others regressive – for example the development of the rules of representation might be progressive with respect to fruitfulness and the development of the rules of scope might be regressive with respect to the same value. In a nutshell, judgements about progress are possible for different kinds of rules with respect to different values within the premises of *one explanatory game.*

Besides, it should be obvious that such judgements about progress can also be made *across different explanatory games*: one can

an action (i.e. relative to the goals of the scientists who accepted or rejected a certain theory). But there is nothing that compels us to make our judgments of the progressiveness of a theory choice depend upon our acquiescence in the aims of science held by those who forged that choice in the first place. Thus we can ask whether Newton's theory of light represented progress over Descartes' optics, without knowing anything about the cognitive aims of Newton or Descartes. Instead, we can (and typically do) make such determinations of progress relative to our own views about the aims and goals of science. [...] All this sounds rather 'whiggish,' and so it should, for when we ask whether science has progressed we are typically asking whether the diachronic development of science has furthered cognitive ends that we deem to be worthy or desirable. Great scientists of the past need not have shared our aims in order for us to ascertain whether their theory choices furthered our cognitive aspirations. For these reasons, a recognition of the fact that aims and values both change does nothing to preclude our use of a robust notion of cognitive scientific progress."

21   Laudan illuminates very nicely this point in the premises of his reticulated picture of science (1984, p. 66): "But if this analysis leaves us with a straightforward procedure for ascertaining whether science has made progress, it forces on us the recognition [...] that progress is always 'progress relative to some set of aims.' Customary usage encourages us to fall into speaking of scientific progress in some absolutist sense; and we are all apt to refer (usually with much hand waving) to scientific progress, without specifying the axiology against which judgments of progress must ultimately be measured. The reticulated picture of science exhibits clearly the dangers of that mistake. Equally, this analysis underscores the fact that a particular bit of science may be progressive (with respect to one set of values) and regressive (with respect to another). There is simply no escape from the fact that determinations of progress must be relativized to a certain set of ends, and that there is no uniquely appropriate set of those ends."

diagnose the progress or regress of one set of rules A (i.e., of representation *or* of inference *or* of scope) used in an explanatory game X in comparison with a set of rules B (i.e., of representation *or* of inference *or* of scope) used in an explanatory game Y with respect to a specific value (i.e., empirical fit, consistency, simplicity, etc.)

Such judgements concerning progress either within one explanatory game or across different explanatory games are relatively unproblematic (though, of course, often contentious) since there is relative clarity concerning the question of what is compared with what (this is always one specific type of rule) and concerning the question of the goal towards which the comparison is being made (this is always one specific value). Such judgements are relatively unproblematic, because they are one-dimensional.

To illustrate the case of progress *with respect to one explanatory game*, let us return to our case study from Economics concerning the explanation of the value of goods (discussed in Chapter 6, Section 6.2). Starting with the *rules of representation*, recall that the classical political economists of the eighteenth century used natural language and numerical examples as the sole means of representation. The marginal economists of the end of the nineteenth and the beginning of the twentieth century introduced the use of the mathematical calculus, more specifically maximization under constraints as a new means of representation. With respect to accuracy, the mathematical calculus is to be judged as better than numerical examples expressed in natural language and there is definitely a case of progress here. With respect to simplicity, however, the mathematical calculus is to be judged as worse than numerical examples expressed in natural language and there is definitely a case of regress here – and so on.

Continuing with the *rules of inference*, recall that the classical political economists used the rules of logic prevailing at that time along with a series of law-like statements, most importantly that the value of a commodity depends on the relative quantity of labour which is necessary for its production. The maximization principle was the most important rule of inference used by the marginal economists for the explanation of the value of goods. With respect to accuracy, the maximization principle is to be judged as better than the law-like statement that the value of a commodity depends on the relative quantity of labour, since it provides enhanced numerical precision. With respect to consistency, the same kind of progress is also apparent. Classical economists derived the prices of products from the "natural" rates of reward of the three factors of production based on three separate principles; that is, wages of labour were determined by the long-run costs of producing the means of subsistence; land rents were determined as a differential surplus over the marginal costs of cultivation; and the rate of profit on capital was regarded as residual. The maximization principle introduced by the marginal economists was a more consistent rule of inference since the prices of products were determined by the application of the same procedure on all factors of production. And so on.

Finally, we turn to the *rules of scope*. The explanation of the value of goods was provided by the classical political economists only in the case of competitive markets in a systematic way, whereas the application of the mathematical calculus and the maximization principle by the marginal economists was also extended to the values of goods under monopolistic conditions, and further to the factors of production. With respect to generality, the rule of scope of the marginal economists, which determined all markets as the appropriate domain of application of the explanatory game, is to be judged as better than the rule used by the classical political economists, which prescribed the application only under competitive conditions. And so on.

Now, let us proceed with the case of judgements about progress in the rules used *across different explanatory games*. Consider the case of the emergence of the universe, cosmogony, in Chapter 7. Recall that the mythical game, the biblical game and the scientific games are structured around the same explanandum, though they differ radically in the other types of constitutive rules, that is, the rules determining what must be taken as given and the rules determining their metaphysical presuppositions. However, since the explanandum remains the same, in principle and in practice the comparability is given. Starting with the *rules of representation*, in both the mythical and the biblical case they consist in narratives expressed in natural language, whereas in the scientific explanatory games they consist in differential equations and computer simulations. With respect to quantitative accuracy the differential equations and computer simulations are to be judged better than the narratives. The same is the case with respect to qualitative accuracy. There is a clear case of progress here. With respect to simplicity, however, the narratives are to be judged as better than the differential equations and computer simulations, if simplicity is specified as cognitive ease.

Continuing with the *rules of inference*, a comparison with respect to consistency is interesting. In the case of both the mythical and the biblical explanatory game, logical consistency among the narratives is not attained, nor is empirical consistency since miracles are explicitly allowed, that is, cases in which conflicting sets of principles are used to accommodate analogous cases. With respect to consistency, thus, there is a clear case of progress attained by the use of logic, statistical inference, etc. in the case of the scientific explanatory game vis-à-vis the other two. Finally, the *rules of scope* seem to be similar in all three cases: the domain of application of all explanations offered is the whole universe.

Though judgements about progress with respect to specific values are possible in the case of one specific set of rules both within the premises of one explanatory game and across different explanatory games, I was cautious not to refer in those one-dimensional cases as judgements of *explanatory progress* per se. The reason is that explanatory progress is the complex outcome of the overall progress of all kinds of rules, that is, of representation, inference and scope. The difficulty in formulating judgements about explanatory progress stems, on the one hand, from the fact that there are three kinds of rules contributing to the solution of explanatory problems and, on the other hand, from the fact that they can be evaluated with respect to different values. Additionally, there might be more specifications of the values to be considered and settled first that could complicate matters even more. For example, accuracy is to be specified first, say as qualitative or quantitative accuracy, simplicity as parsimony or cognitive ease, etc. Such decisions are, thus, more complex since they are the function of the multi-dimensional evaluation of different rules. They are also of increased complexity because of the fact that some rules are more of a theoretical nature whereas other rules incorporate procedural knowledge.

However, complex decisions are not impossible – they are more difficult to make since there are more factors to be taken into consideration, but they can and must be made nevertheless. Since they are naturally more prone to error than the simpler one-dimensional decisions about the progress of distinct sets of rules, the importance of formal and informal institutions in which the respective explanatory games are embedded becomes obvious: they can offer the means of correcting the permanent errors that we are always going to make when making such complex decisions about explanatory progress.

And naturally there are two kinds of explanatory progress to be distinguished here: *local explanatory progress* when judging the situation with respect to one explanatory game and *global explanatory progress* when judging the situation with respect to different explanatory games. Judgements about local explanatory progress are somehow less complex since one has to decide whether the simultaneous change of some rules with respect to some values has to count as *explanatory progress* or not. Such judgements are not to be made on the basis of the application of some simple criterion of explanatory progress to be formulated and offered here as a definitive proposal. Depending on the context there might be clear-cut cases where there is simultaneous progress of all rules with respect to all values in which case a local explanatory progress is clearly discernible. And there might be hard cases where there is a variety of progress and regress across different dimensions where judgements about explanatory progress involve the weighing among values, their ordering and the conscious disregard of some of them before proceeding to an informed choice.

The decisions regarding *global explanatory progress* across different explanatory games are of obvious importance. Here the complexity of the situation is of course even greater: the sole condition of comparability is that the constitutive rules determining what counts as an explanandum are the same across different explanatory games.[22] This increases the complexity of the situation. The reason for this is that even though a comparative evaluation with respect to the rules of representation, inference and scope might turn out to be unproblematic, there are always differences with respect to the other constitutive rules, that is, with respect to the rules determining what must be taken as given and the rules determining the metaphysical presuppositions. To employ the example of the explanation of cosmogony again, consider the comparative evaluation between the biblical and the scientific explanatory games. Since the rules of scope are similar or identical (i.e., since the whole universe is the domain of application), an evaluation concerns the rules of representation and the rules of inference. With respect to almost all epistemic values, the rules of representation and the rules of inference of the scientific explanatory game fare better than those of the biblical game: accuracy, consistency, fruitfulness, etc. Only with respect to simplicity are the rules of representation and inference of the biblical game to be judged better. The complexity of the decision regarding the superiority of the one game over the other is high, thus, due to the need of weighing the importance of different values and the like.

Instead of focusing on (formal or even syntactic criteria) to adjudicate these cases, the explanatory methodology of the genre that I am proposing here requires another solution, one which has to do with the conditions that should prevail in order for a critical debate

---

[22] There might be some evident cases where judgements about global explanatory progress are easy to make – say when comparing explanatory activities of children and adults aimed at the same explanandum– but these cases are bound to remain exceptional and very often simply uninteresting.

to take place informing the making of such judgements: it is the institutional framework, both formal and informal, in which the respective explanatory games are embedded that plays the – literally – decisive role here. *Global explanatory progress* is tied to the *procedural rationality* that I suggest, which does not prescribe the content of the decisions that are ultimately made by the participants of the explanatory games themselves, but highlights instead the *individual* and *collective* conditions that must prevail in order for these decisions and judgements to be adequate with respect to the goals and values that they are aiming at. The *individual conditions* are relevant to helping secure the possibility of constructing multiple alternative judgements which function as the diagnosis of a global scientific progress and the *collective conditions* are relevant to helping secure the possibility of their criticism. In the end this is possible only through the existence of an appropriate institutional framework that provides the safeguards that enable the prevailing of these conditions, most importantly the freedom of expression.

12 Epilogue

One way to defend explanatory pluralism is to view the myriad of explanatory activities undertaken by scientists and ordinary people in terms of explanatory games. In social life and in different domains of science different explanatory games are played, and the philosophical project consists in describing and normatively appraising the rules that constitute these games. This project is fundamentally liberal, in the sense that participants and non-participants to the game alike engage in the critical discussion and revision of the rules or to put it in other terms, the project is fundamentally naturalistic – philosophers and scientists equally take part in it.

The explanatory enterprise being a process that is developing, all we can do is to point to the constraints of this process, the rules of the game, but we can never work out the necessary and sufficient conditions of what should be called an "explanation".[1] *What is an explanation?* is a question that cannot be answered by a single philosophical account of explanation, but it is up to the various explanatory games to come to terms with this. The traditional theory suggests that we explain if and only if we use a theory of a certain sort (e.g., mechanistic, unficationist, manipulationist). According to my suggestion we explain if and only if we use a theory insofar as this theory is embedded in a certain explanatory game, where the "explanatory" is defined as a network of certain rule-guided practices.

A virtue of my proposal consists in the identifiability of an explanatory game, thus allowing for the accurate description, comparison and evaluation of the rules and the application of the explanatory

---

[1] See Nietzsche's (1887) commentary (*Zur Genealogie der Moral* II: 13): "[...] alle Begriffe, in denen sich ein ganzer Prozess semiotisch zusammenfasst, entziehen sich der Definition; definierbar ist nur das, was keine Geschichte hat".

methodology along the lines that I have outlined. The tractability of an explanatory game seems to be higher than all alternative approaches which have dominated the discussion: paradigms (Kuhn), research programmes (Lakatos), research traditions (Laudan), consensus practices (Kitcher) and relativized constitutive a priori principles (Friedman). The main reason lies in the clearer, more specific focus: rather than science in general, it is explanatory activities that have been the topic of the present inquiry. Though these practices do not exhaust science, they lie at its heart, and certainly make up the core of theoretical science.

However, the main virtue of my approach certainly lies in the minimalist metaphysics to which it is committed. Many realist philosophers of science do accept that there are several values underlying theory choice and that these values need to be traded off against each other. So, they admit that it is virtually impossible to characterize in a uniform way the complexities of what counts as a choice-worthy theory. Since these realists are already happy to be pluralists about theory choice, they should – by analogy – be also open to pluralism about explanation, something they are traditionally not open to. Constructive and other empiricists who insist that our theories are only means to save the phenomena and are hesitant to make claims about the ontological structure of reality can also embrace a pluralistic stance to explanation which is not committed to any detailed theses regarding ontology.

The explanations generated while an explanatory game is played can be taken, hence, to delineate the objective order of dependencies in nature as the scientific realist insists or for someone who does not believe in a mind-independent order of nature, the explanations can be taken to present part of a system of organizing the phenomena. Explanatory progress takes place when we advance our true or truth-like knowledge about the way natural phenomena depend on each other or when we increase the degree of organizing various areas of our experience. Such ecumenism being closer to the explanatory practices both of scientists and ordinary people, it does not ban metaphysical questions from the broad picture of the approach, but it designates their appropriate locus in the debates about the metaphysical presuppositions of the explanatory games which partly constitute them. (This is the reason that I have also consistently avoided refering to truth as a value; from the perspective of participants in a religious explanatory game who commonly reserve the notion of truth to describe the main property of a deity, debates about versions of the correspondence theory of truth must seem as artificial puzzles dealing with aesthetically pleasing irrelevancies.[2])

And there are certainly a series of weaknesses and limitations to my approach; critical discussion will show where they lie.

---

[2]   In 1992, the 400th anniversary of Galileo's appointment at the University of Padua, Pope John Paul II wrote to the Rector of the University of Padua, acknowledging that the Catholic Church was in error in that controversy: "The common vocation of both scientists and theologists is to contribute to the best knowledge of the Truth." (November 26, 1992, Il Gazzettino 5– Reference from: Thiene 1997, p. 88).

References

Abney, Drew H., Rick Dale, Jeff Yoshimi, Chris T. Kello, Kristian Tylén, and Ricardo Fusaroli (2014): "Joint Perceptual Decision-Making: A Case Study in Explanatory Pluralism", *Frontiers in Psychology*, vol. 5, pp. 1–12.

Achinstein, Peter (1983): *The Nature of Explanation*, New York: Oxford University Press.

Acierno, Louis (1994): *The History of Cardiology*, London and New York: The Parthenon Publishing Group.

Adams, Marcus (2009): "Empirical Evidence and the Knowledge-that/Knowledge-how Distinction", *Synthese*, vol. 170, pp. 97–114.

Aird, W.C. (2011): "Discovery of the Cardiovascular System: From Galen to William Harvey", *Journal of Thrombosis and Haemostasis*, vol. 9 (Suppl. 1), pp. 118–129.

Albert, Hans (1968/1985): *Treatise on Critical Reason*, Princeton: Princeton University Press.

Albert, Hans (1978): *Traktat über Rationale Praxis*, Tübingen: J.C.B. Mohr (Paul Siebeck).

Albert, Hans (1982): *Die Wissenschaft und die Fehlbarkeit der Vernunft*, Tübingen: J.C.B. Mohr (Paul Siebeck).

Albert, Hans (1987): *Kritik der Reinen Erkenntnislehre*, Tübingen: J.C.B. Mohr (Paul Siebeck).

Alt, James (2009): "Comment: Conditional Knowledge: An Oxymoron?" in C. Mantzavinos (ed.): *Philosophy of the Social Sciences. Philosophical Theory and Scientific Practice*, Cambridge: Cambridge University Press, pp. 146–153.

Anderson, John R. (2010): *Cognitive Psychology and Its Implications*, 7th edition, New York: Worth Publishers.

Arabatzis, Theodore (2006): "On the Inextricability of the Context of Discovery and the Context of Justification", in Jutta Schickore and Friedrich Steinle (eds.): *Revisiting Discovery and Justification*, Dordrecht: Springer, pp. 215–230.

Axelrod, Robert (1986): "An Evolutionary Approach to Norms", *American Political Science Review*, vol. 80, pp. 1095–1111.

Aulus Cornelius Celsius (1935): *De Medicina (On Medicine)*, vol. 1, Loeb Classical Library edition, translated by W.G. Spencer, Cambridge: Harvard University Press.

Baker, Alan (2012): "Science-Driven Mathematical Explanations", *Mind*, vol. 121, pp. 243–267.

Bandura, Albert (1986): *Social Foundations of Thought and Action: A Social Cognitive Theory*, Englewood Cliffs, NJ: Prentice-Hall.

Bargh, John A. and Tanya L. Chartrand (1999): "The Unbearable Automaticity of Being", *American Psychologist*, vol. 54, pp. 462–479.

Barnes, Barry, David Bloor, and John Henry (1996): *Scientific Knowledge. A Sociological Analysis*, Chicago: The University of Chicago Press.

Bartelborth, Thomas (2002): "Explanatory Unification", *Synthese*, vol. 130, pp. 91–107.

Bartelborth, Thomas (2007): *Erklären*, Berlin and New York: de Gruyter.

Bartlett, Frederic (1932): *Remembering*, Cambridge: Cambridge University Press.

Batterman, Robert W. (2010): "On the Explanatory Role of Mathematics in Empirical Science", *British Journal for the Philosophy of Science*, vol. 61, pp. 1–25.

Bechtel, William (2006): *Discovering Cell Mechanisms*, Cambridge: Cambridge University Press.

Bechtel, William (2008): *Mental Mechanisms: Philosophical Perspectives on Cognitive Neuroscience*, London: Routledge.

Bechtel, William (2011): "Mechanism and Biological Explanation", *Philosophy of Science*, vol. 78, pp. 533–557.

Bechtel, William and Adele Abrahamsen (2005): "Explanation: A Mechanist Alternative", *Studies in History and Philosophy of Biological and Biomedical Sciences*, vol. 36, pp. 421–441.

Becker, Gary (1976): *The Economic Approach to Human Behavior*, Chicago and London: The University of Chicago Press.

Becker, Gary and Richard Posner (2009): *Uncommon Sense: Economic Insights from Marriage to Terrorism*, Chicago and London: The University of Chicago Press.

Bernholz, Peter, Manfred E. Streit and Roland Vaubel (1998): *Political Competition, Innovation and Growth*, Berlin: Springer.

Bible, Old Testament.

Binmore, Ken and Larry Samuelson (1994): "An Economist's Perspective on the Evolution of Norms", *Journal of Institutional and Theoretical Economics*, vol. 150, pp. 45–63.

Bird, Alexander (2005): "Explanation and Metaphysics", *Synthese*, vol. 143, pp. 89–107.

Blaug, Marc (1997): *Economic Theory in Retrospect*, 5th edition, Cambridge: Cambridge University Press.

Bogen, Jim and Peter Machamer (2011): "Mechanistic Information and Causal Continuity", in Phyllis McKay Illari, Federica Russo, and Jon Williamson (eds.): *Causality in the Sciences*, Oxford: Oxford University Press, pp. 845–864.

Boudon, Raymond (1974): *Education, Opportunity and Social Inequlity*, New York: Wiley.

Boyer, Pascal (2000): "Evolution of the Modern Mind and the Origins of Culture: Religious Concepts as a Limiting-Case", in Peter Carruthers and Andrew Chamberlain (eds.): *Evolution and the Human Mind. Modularity, Language and Meta-Cognition*, Cambridge: Cambridge University Press, pp. 93–112.

Boyle, Robert (1688): *A Disquisition About the Final Causes of Natural Things: Wherein It is Inquired, Whether, and, (if at all) with What Cautions a Naturalist Should Admit Them?* London: John Caplot.

Brain, Peter (1986): *Galen on Bloodletting*, Cambridge: Cambridge University Press.

Braine, David (1972): "Varieties of Necessity", *Supplementary Proceedings of the Aristotelian Society*, vol. 46, pp. 139–170.

Braithwaite, Richard B. (1953): *Scientific Explanation. A Study of the Function of Theory, Probability and Law in Science*, Cambridge: Cambridge University Press.

Brandt, Karl (1993): *Geschichte der Deutschen Volkswirtschaftslehre*, vol. 2, Freiburg i. Br.: Rudolf Haufe Verlag.

Brennan, Geoffrey and James Buchanan (1985): *The Reason of Rules – Constitutional Political Economy*, Cambridge: Cambridge University Press.

Brewer, William F., Clark A. Chinn, and Ala Samarapungavan (2000): "Explanation in Scientists and Children", in Frank C. Keil and Robert A. Wilson (eds.): *Explanation and Cognition*, Cambridge, MA: The MIT Press, pp. 279–298.

Bromberger, Sylvain (1966): "Why-Questions", in Robert Colodny (ed.): *Mind and Cosmos*, Pittsburgh: University of Pittsburgh Press, pp. 86–111.

Brown, Matthew J. (2013): "Values in Science beyond Underdetermination and Inductive Risk", *Philosophy of Science*, Vol. 80, pp. 829–839.

Buchanan, James (1975): *The Limits of Liberty. Between Anarchy and Leviathan*, Chicago: The University of Chicago Press.

Buchanan, James and Gordon Tullock (1962): *The Calculus of Consent – Logical Foundations of Constitutional Democracy*, Ann Arbor: University of Michigan Press.

Bueno, Otávio and Steven French (2012): "Can Mathematics Explain Physical Phenomena?" *British Journal for Philosophy of Science*, vol. 63, pp. 85–113.

Bueter, Anke (2015): "The Irreducibility of Value-Freedom to Theory Assessment", *Studies in History and philosophy of Science*, Vol. 49, pp. 18–26.

Burkert, Walter (1999): "The Logic of Cosmogony", in Richard Buxton (ed.): *From Myth to Reason? Studies in the Development of Greek Thought*, Oxford: Oxford University Press, pp. 87–106.

Butterfield, Herbert (1957): *The Origins of Modern Science 1300–1800*, rev. edition, New York: Free Press.

Campbell, Donald T. (1965): "Variation and Selective Retention in Socio-Cultural Evolution", in Herbert R. Barringer, George I. Blanksten, and Raymond W. Mack (eds.): *Social Change in Developing Areas*, Cambridge, MA: Schenkman, pp. 19–49.

Campbell, Donald T. (1974/1987): "Evolutionary Epistemology", originally published in Paul A. Schlipp (ed.): *The Philosophy of Karl Popper*, La Salle, IL: Open Court, pp. 413–463, and reprinted in Gerard Radnitzky and W.W. Bartley III (eds.) (1987): *Evolutionary Epistemology, Rationality and the Sociology of Knowledge*, La Salle, IL: Open Court, pp. 48–114.

Carey, Susan (1985): *Conceptual Change in Childhood*, Cambridge, MA: The MIT Press.

Carnap, Rudolf (1932): "Überwindung der Metaphysik durch logische Analyse der Sprache", *Erkenntnis*, vol. 2, pp. 219–241.

Carrier, Martin (2013): "Values and Objectivity in Science: Value-Ladeness, Pluralism and the Epistemic Attitude", *Science & Education*, vol. 22, pp. 2547–2568.

Carruthers, Peter (2002): "The Roots of Scientific Reasoning: Infancy, Modularity and the Art of Trucking", in Peter Carruthers, Stephen Stich, and Michael Siegal (eds.): *The Cognitive Basis of Science*, Cambridge: Cambridge University Press, pp. 73–95.

Cartwright, Nancy (1999): *The Dappled World. A Study of the Boundaries of Science*, Cambridge: Cambridge University Press.

Cartwright, Nancy (2007): *Hunting Causes and Using Them*, Cambridge: Cambridge University Press.

Celsus, *On Medicine, Prooemium*, Loeb Classical Library edition, Cambridge, MA: Harvard University Press.

Cesalpino, Andrea (1571): *Peripateticarum Questionum, Libri Quinque*, Venice: Apud Iuntas.

Cesalpino, Andrea (1593): *Quaestionum Medicarum, Libri II*, Venice: Apud Iuntas.

Cesalpino, Andrea (1606): *Praxis Universae Artis Medicae*, Treviso: Sumptibus Robert Meietti.

Chang, Hasok (2012): *Is Water H₂O? Evidence, Realism and Pluralism*, Berlin and New York: Springer.

Clark, Matthew (2012): *Exploring Greek Myth*, Malden: Wiley-Blackwell.

Cohen, Neal J. and Larry R. Squire (1980): "Preserved Learning and Retention of Pattern-Analyzing Skill in Amnesia: Dissociation of Knowing How and Knowing That", *Science*, vol. 210, pp. 207–210.

Coleman, James S. (1990a): *Foundations of Social Theory*, Cambridge, MA: Harvard University Press.

Coleman, James S. (1990b): "The Emergence of Norms", in Michael Hechter, Karl-Dieter Opp, and Reinhard Wippler (eds.): *Social Institutions*, Berlin and New York: Walter de Gruyter, pp. 35–59.

Colombo, Matteo and Stephan Hartmann (2015): "Bayesian Cognitive Science, Unification, and Explanation", *British Journal for the Philosophy of Science* (Advanced Access) published September 30, 2015.

Colombo, Matteo, Stephan Hartmann, and Robert van Iersel (2015): "Models, Mechanisms, and Coherence", *British Journal for the Philosophy of Science*, vol. 66, pp. 181–212.

Colombo, Realdo (1559): *De Re Anatomica*, Libri XV, Venice: Nicolai Beuilacquae.

Coppola, Edward D. (1957): "The Discovery of the Pulmonary Circulation: A New Approach", *Bulletin of the History of Medicine*, vol. 21, pp. 44–77.

Cosmides, Leda and Tooby John (eds.) (1992): *The Adapted Mind*, Oxford: Oxford University Press.

Cournot, Augustin (1838): *Recherches sur les Principes Mathématiques de la Théorie des Richesses*, Paris: Chez L. Hachette.

Craver, Carl (2007): *Explaining the Brain. Mechanisms and the Mosaic Unity of Neuroscience*, Oxford: Oxford University Press.

Crupi, Vincenzo and Katya Tentori (2012): "A Second Look at the Logic of Explanatory Power (with Two Novel Representation Theorems)", *Philosophy of Science*, vol. 79, pp. 365–385.

Currie, Adrian Mitchell (2014): "Narratives, Mechanisms and Progress in Historical Science", *Synthese*, vol. 191, pp. 1163–1183.

D'Andrade, Roy (1995): *The Development of Cognitive Anthropology*, Cambridge: Cambridge University Press.

Darden, Lindley (2006): *Reasoning in Biological Discoveries. Essays in Mechanisms, Interfield Relations and Anomaly Resolution*, Cambridge: Cambridge University Press.

Daston, Lorraine and Peter Galison (2010): *Objectivity*, New York: Zone Books.

Davidson, Janet E. and Robert J. Sternberg (eds.) (2003): *The Psychology of Problem Solving*, Cambridge: Cambridge University Press.

Davies, Martin (2015): "Knowledge (Explicit, Implicit and Tacit): Philosophical Aspects", *International Encyclopedia of the Social and Behavioral Sciences*, 2nd edition, vol. 13, London: Elsevier, pp. 74–90.

de Regt, Henk W. (2006): "Wesley Salmon' s Complementarity Thesis: Causalism and Unificationism Reconciled?" *International Studies in the Philosophy of Science*, vol. 20, pp. 129–147.

Debru, Armelle (2008): "Anatomy", in Robert J. Hankinson (ed.): *The Cambridge Companion to Galen*, Cambridge: Cambridge University Press, pp. 263–282.

Demeulenaere, Pierre (ed.) (2011): *Analytical Sociology and Social Mechanisms*, Cambridge: Cambridge University Press.

Denzau, Arthur and Douglass C. North (1994): "Shared Mental Models: Ideologies and Institutions", *Kyklos*, vol. 47, pp. 3–31.

Devitt, Michael (2011): "Methodology and the Nature of Knowing How", *Journal of Philosophy*, vol. 108, pp. 205–218.

Díez, José, Khalifa Kareem, and Leuridan Bert (2013): "General Theories of Explanation: Buyer Beware", *Synthese*, vol. 190, pp. 379–396.

Donald, Merlin (1991): *The Origins of the Modern Mind. Three Stages in the Evolution of Culture and Cognition*, Cambridge, MA: Harvard University Press.

Douglas, Heather (2013): "The Value of Cognitive Values", *Philosophy of Science*, vol. 80, pp. 796–806.

Dowe, Philip (2000): *Physical Causation*, Cambridge: Cambridge University Press.

Downes, Anthony (1957): *An Economic Theory of Democracy*, New York: Harper.

Dupré, John (1993): *The Disorder of Things*, Cambridge, MA: Harvard University Press.

Eknoyan, Garabed and Natale G. De Santo (1997): "Realdo Colombo (1516–1559). A Reappraisal", *American Journal of Nephrology*, vol. 17, pp. 261–268.

Elster, Jon (1989a): *The Cement of Society. A Study of Social Order*, Cambridge: Cambridge University Press.

Elster, Jon (1989b): "Social Norms and Economic Theory", *Journal of Economic Perspectives*, vol. 3, pp. 99–107.

Elster, Jon (2007): *Explaining Social Behavior*, Cambridge: Cambridge University Press.

Epictetus (1916): *Epicteti Dissertationes ab Arriano digestae*, ed. Heinrich Schenkl, Leipzig: B.G.Teubner.

Fogan, Melinda Bonnie (2015): "Collaborative Explanation and Biological Mechanism", *Studies in History and Philosophy of Science*, vol. 52, pp. 67–78.

Fancy, Nahyan (2013): *Science and Religion in Mamluk Egypt: Ibn al-Nafis, Pulmonary Transit and Bodily Resurrection*, London and New York: Routledge.

Faucher, Luc, Ron Mallon, Daniel Nazer, Shaun Nichols, Aaron Ruby, Stephen Stich, and Jonathan Weinberg (2002): "The Baby in the Lab-Coat: Why Child Development Is Not an Adequate Model for Understanding the Development of Science", in Peter Carruthers, Stephen Stich, and Michael Siegal (eds.): *The Cognitive Basis of Science*, Cambridge, MA: The MIT Press, pp. 335–362.

Faye, Jan (2012): *After Postmodernism. A Naturalistic Reconstruction of the Humanities*, Houndmills: Palgrave Macmillan.

Faye, Jan (2014): *The Nature of Scientific Thinking. On Interpretation, Explanation and Understanding*, Houndmills: Palgrave Macmillan.

Ferguson, Adam (1792): *Principles of Moral and Political Science*, Edinburgh: W. Creech.

Feyerabend, Paul (1975/2010): *Against Method*, 4th edition, London and New York: Verso.

Files, Craig (1996): "Goodman's Rejection of Resemblance", *British Journal of Aesthetics*, vol. 36, pp. 398–412.

Fisher, George Jackson (1880): "Hieronymus Fabricius ab Acquapendente", *Annals of Anatomy and Surgery*, pp. 373–379.

Fodor, Jerry (1983): *The Modularity of Mind*, Cambridge, MA: The MIT Press.

Franklin-Hall, Laura R. (2014): "High-Level Explanation and the Interventionist's 'Variables Problem'", *British Journal for the Philosophy of Science* (Advance Access), published December 18, 2014.

Frede, Michael (ed.) (1985): *Galen: Three Treatises on the Nature of Science*, Indianapolis: Hackett Publishing.

Frede, Michael (1987): "On the Method of the So-Called Methodical School of Medicine", in Michael Frede (ed.): *Essays in Ancient Philosophy*, Oxford: Oxford University Press, pp. 261–278 [originally published in Barnes, Jonathan, Jacques Brunschwig, Myles Burnyeat, and Malcolm Schofield (eds.) (1982): *Science and Speculation*, Cambridge: Cambridge University Press, pp. 1–23].

Frege, Gottlob (1893/1998): *Grundgesetze der Arithmetik*, Hildesheim, Zürich and New York: Georg Olms Verlag.

Friedman, Michael (1974): "Explanation and Scientific Understanding", *Journal of Philosophy*, vol. 71, pp. 5–19.

Friedman, Michael (2001): *Dynamics of Reason*, Stanford, CA: CSLI Publications.

Friedman, Michael (2002): "Kant, Kuhn and the Rationality of Science", *Philosophy of Science*, vol. 69, pp. 171–190.

Friedman, Michael (2011): "Extending the Dynamics of Reason", *Erkenntnis*, vol. 75, pp. 431–444.

Friedman, Michael (2012): "Reconsidering the Dynamics of Reason: Response to Ferrari Mormann, Nordmann, and Uebel", *Studies in History and Philosophy of Science*, vol. 43, pp. 47–53.

Fries, Jakob Friedrich (1828–1831): *Neue oder anthropologische Kritik der Vernunft*, vols. 1–3, 2nd edition, Heidelberg: Christian Friedrich Winter.

Frigg, Roman and Matthew Hunter (eds.) (2010): *Beyond Mimesis and Convention. Representation in Art and Science*, Dordrecht, Heidelberg, London and New York: Springer.

Fullbrook, Edward (ed.) (2008): *Ontology and Economics: Tony Lawson and His Critiques*, London: Routledge.

Furley, David. J. and Wilkie, J.S. (1984): *Galen on Respiration and the Arteries*, Princeton: Princeton University Press.

Gentner, Dedre and Donald R. Gentner (1983): "Flowing Waters or Teeming Crowds: Mental Models of Electricity", in Dedre Gentner and Albert Stevens (eds.): *Mental Models*, Hillsdale, NJ: Lawrence Erlbaum, pp. 99–129.

Gervais, Raoul (2014): "A Framework for Inter-Level Explanations: Outlines for a New Explanatory Pluralism", *Studies in History and Philosophy of Science*, vol. 48, pp. 1–9.

Giere, Ronald (1996): "The Scientist as Adult", *Philosophy of Science*, vol. 63, pp. 538–541.

Giere, Ronald (2004): "How Models are Used to Represent Reality", *Philosophy of Science*, vol. 71, pp. 742–752.

Giere, Ronald (2006): *Scientific Perspectivism*, Chicago: The University of Chicago Press.

Gigerenzer, Gerd (2008): "Why Heuristics Work", *Perspectives on Psychological Science*, vol. 3, pp. 20–29.

Gigerenzer, Gerd, Peter Todd, and the ABC Research Group (1999): *Simple Heuristics that Make Us Smart*, Oxford: Oxford University Press.

Gigerenzer, Gerd, Ralph Hertwig, and Thorsten Pachur (eds.) (2011): *Heuristics: The Foundations of Adaptive Behavior*, Oxford: Oxford University Press.

Gijsbers, Victor (2007): "Why Unification is Neither Necessary Nor Sufficient for Explanation", *Philosophy of Science*, vol. 74: pp. 481–500.

Glennan, Stuart (2002): "Rethinking Mechanistic Explanation", *Philosophy of Science*, vol. 69, pp. 342–353.

Glennan, Stuart (2010): "Ephemeral Mechanisms and Historical Explanation", *Erkenntnis*, vol. 72, pp. 251–266.

Glymour, Clark (2015): "Probability and the Explanatory Virtues", *British Journal for the Philosophy of Science*, vol. 66, pp. 591–604.

Godfrey-Smith, Peter (2003): *Theory and Reality*, Chicago: The University of Chicago Press.

Goodman, Nelson (1976): *Languages of Art. An Approach to a Theory of Symbols*, Indiana: Hackett Publishing Company.

Gopnik, Alison (1996a): "The Scientist as Child", *Philosophy of Science*, vol. 63, pp. 485–514.

Gopnik, Alison (1996b): "Reply to Commentators", *Philosophy of Science*, vol. 63, pp. 552–561.

Gopnik, Alison (2000): "Explanation as Orgasm and the Drive for Causal Knowledge: The Function, Evolution, and Phenomenology of the Theory

Formation System", in Frank C. Keil and Robert A. Wilson (eds.): *Explanation and Cognition*, Cambridge, MA: The MIT Press, pp. 299–323.

Gopnik, Alison (2012): "Scientific Thinking in Young Children: Theoretical Advances, Empirical Research, and Policy Implications", *Science*, vol. 337 (no. 6102), pp. 1623–1627.

Gopnik, Alison and Andrew Meltzhoff (1998): *Words, Thoughts, and Theories*, Cambridge, MA: The MIT Press.

Gossen, Hermann Heinrich (1854): *Entwicklung der Gesetze des menschlichen Verkehrs, und der daraus fließenden Regeln für menschliches Handeln*, Braunschweig: Friedrich Vieweg & Sohn.

Grantham, Todd A. (1999): "Explanatory Pluralism in Paleobiology", *Philosophy of Science*, vol. 66, pp. S223–S236.

Grantham, Todd A. (2004): "Conceptualizing the (Dis)unity of Sciences", *Philosophy of Science*, vol. 71, pp. 133–155.

Greenberg, Gabriel (2013): "Beyond Resemblance", *Philosophical Review*, vol. 122, pp. 215–287.

Grimm, Stephen R. (2008): "Explanatory Inquiry and the Need for Explanation", *British Journal for the Philosophy of Science*, vol. 59, pp. 481–497.

Hacking, Ian (2002): *Historical Ontology*, Cambridge, MA: Harvard University Press.

Halonen, Ilpo and Jaako Hintikka (2005): "Toward a Theory of the Process of Explanation", *Synthese*, vol. 143, pp. 5–61.

Hankinson, Robert J. (2008): "The Man and His Work", in Robert J. Hankinson (ed.): *The Cambridge Companion to Galen*, Cambridge: Cambridge University Press, pp. 1–33.

Harbecke, Jens (2010): "Mechanistic Constitution in Neurobiological Explanations", *International Studies in the Philosophy of Science*, vol. 24 (3), pp. 267–285.

Harman, Gilbert (1965): "The Inference to the Best Explanation", *Philosophical Review*, pp. 88–95.

Harvey, William (1628/1989): *Exercitatio Anatomica de Motu Cordis et Sanguinis in Animalibus*, English translation by Robert Willis in The Works of William Harvey, Philadelphia: University of Pennsylvania Press.

Hayek, Friedrich A. von (1955): "Degrees of Explanation", *British Journal for the Philosophy of Science*, vol. 6, pp. 209–225.

Hayek, Friedrich A. von (1967): "The Theory of Complex Phenomena", in Friedrich A. von Hayek: *Studies in Philosophy, Politics, and Economics*, London: Routledge, pp. 22–42.

Hayek, Friedrich A. von (1982): *Law, Legislation and Liberty*, 3 volumes, London: Routledge and Kegan Paul.

Hedström, Peter and Richard Swedberg (eds.) (1996): *Social Mechanisms. An Analytical Approach to Social Theory*, Cambridge: Cambridge University Press.

Helmreich, G. (1907–1909): *Galeni. De Usu Partium Libri XVII*, Leipzig: Teubner.

Hempel, Carl G. (1965): *Aspects of Scientific Explanation and Other Essays in the Philosophy of Science*, New York: The Free Press.

Hempel, Carl G. and Paul Oppenheim (1948): "Studies in the Logic of Explanation", *Philosophy of Science*, vol. 15, 135–175 [Reprinted in Hempel (1965)].

Herfeld, Catherine (2012): *The Many Faces of Rational Choice Theory*, Ph.D. Dissertation, Witten/Herdecke University.

Herrick, James Bryan (1942): *A Short History of Cardiology*, Springfield, IL: Charles C. Thomas.

Hesiod: *Theogony*, Penguin Classics, London.

Hey, Spencer Phillips (2014): "Heuristics and Meta-Heuristics in Scientific Judgment", *British Journal for the Philosophy of Science* (Advance Access), published December 24, 2014.

Hieronymus Fabricius ab Acquapendendte (1603): *De Venarum Ostiolis*, facsimile edition, English translation by K.J. Franklin, D.M., 1933, Springfeld, IL, and Baltimore, MD: Charles C. Thomas.

Holland, John H., Keith J. Holyoak, Richard Nisbett, and Paul Thagard (1986): *Induction: Processes of Inference, Learning and Discovery*, Cambridge, MA: The MIT Press.

Hull, David (1988): *Science as a Process*, Chicago: The University of Chicago Press.

Humphreys, Paul. (1981): "Aleatory Explanation", *Synthese*, 48, pp. 225–232.

Humphreys, Paul (1989): *The Chances of Explanation*, Princeton: Princeton University Press.

Hutchins, Edwin and Hazlehurst Brian (1992): "Learning in the Culture Process", in Christopher G. Langdon, Charles Taylor, J. Doyne Farmer, and Rasmussen Steen (eds.): *Artificial Life II*, Redwood City, CA: Addison-Wesley, pp. 689–706.

Iskandar, A.Z. (ed.) (1988): *Galeni De Optimo Medico Cognoscendo*, Corpus Medicorum Graecorum, Suppl. Or. IV [Arabic], Berlin: Berlin-Bradenburgische Akademie der Wissenschaften.

Jackson, Frank and Philip Pettit (1992): "In Defense of Explanatory Ecumenism", *Economics and Philosophy*, vol. 8, pp. 1–21.

Jaffé, William (ed.) (1965): *Correspondence of Léon Walras and Related Papers*, vol. II, Amsterdam: North-Holland Publishing Company.

Jansson, Lina (2014): "Causal Theories of Explanation and the Challenge of Explanatory Disagreement", *Philosophy of Science*, vol. 81, pp. 332–348.

Jevons, William Stanley (1871): *The Theory of Political Economy*, London and New York: MacMillan & Co.

Johnson-Laird, Philip N. (1983): *Mental Models*, Cambridge, MA: Harvard University Press.

Johnson-Laird, Philip N. (2006): *How We Reason*, Oxford: Oxford University Press.

Jones, Eric L. (2003): *The European Miracle*, 3rd edition, Cambridge: Cambridge University Press.

Keil, Frank C. and Robert A. Wilson (2000): "Explaining Explanation", in Frank C. Keil and Robert A. Wilson (eds.): *Explanation and Cognition*, Cambridge, MA: The MIT Press, pp. 1–18.

Kellert, Stephen H., Helen E. Longino, and C. Kenneth Waters (2006): "Introduction: The Pluralist Stance", in Stephen H. Kellert, Helen E. Longino, and C. Kenneth Waters (eds.): *Scientific Pluralism*, Minneapolis and London: University of Minnesota Press, pp. vii–xxix.

Keuth, Herbert (1993): *Erkenntnis oder Entscheidung. Zur Kritik der kritischen Theorie*, J.C.B. Mohr (Paul Siebeck).

Key, Jack D., Thomas E. Keys, and John A. Callahan (1979): "Historical Development of Concept of Blood Circulation", *The American Journal of Cardiology*, vol. 43, pp. 1026–1032.

Khalifa, Kareem (2012): "Inaugurating Understanding or Repackaging Explanation?" *Philosophy of Science*, vol. 79, pp. 15–37.

Khan, Ijaz A., Samantapudi K. Daya, and Ramesh M. Gowda (2005): "Evolution of the Theory of Circulation", *International Journal of Cardiology*, vol. 98, pp. 519–521.

Kitcher, Philip (1981): "Explanatory Unification", *Philosophy of Science*, vol. 48, pp. 251–281.

Kitcher, Philip (1985): "Two Approaches to Explanation", *Journal of Philosophy*, vol. 82, pp. 632–639.

Kitcher, Philip (1989): "Explanatory Unification and the Causal Structure of the World", in Philip Kitcher and Wesley Salmon (eds.): *Scientific Explanation*, volume 13 of Minnesota Studies in the Philosophy of Science Minneapolis: University of Minnesota Press, pp. 410–505.

Kitcher, Philip (1993): *The Advancement of Science*, Oxford: Oxford University Press.

Kitcher, Philip (2001): *Science, Truth, and Democracy*, Oxford: Oxford University Press.

Kitcher, Philip (2011a): *The Ethical Project*, Cambridge, MA: Harvard University Press.

Kitcher, Philip (2011b): *Science in a Democratic Society*, New York: Prometheus Books.

Kitcher, Philip and Achille Varzi (2000): "Some Pictures Are Worth 2^{N_0} Sentences", *Philosophy*, vol. 75, pp. 377–381.

Knight, Jack (2009): "Causal Mechanisms and Generalizations", in C. Mantzavinos (ed.): *Philosophy of the Social Sciences. Philosophical Theory and Scientific Practice*, Cambridge: Cambridge University Press, pp. 179–184.

Koertge, Noretta (1992): "Explanation and Its Problems", *British Journal for the Philosophy of Science*, vol. 43, pp. 85–98.

Koethe, John (2002): "Stanley and Williamson on Knowing How", *The Journal of Philosophy*, vol. 99, pp. 325–328.

Kühn, Carl Gottlob (ed.) (1821–1833): *Claudii Galeni Opera Omnia*, Leipzig: C. Cnobloch, rpt. Hildesheim: Georg Olms, 1964–1965.

Kuhn, Thomas (1962/1970): *The Structure of Scientific Revolutions*, second enlarged edition, Chicago: Chicago University Press.

Kuhn, Thomas (1977): "Objectivity, Value Judgment, and Theory Choice", in Thomas Kuhn: *The Essential Tension*, Chicago: The University of Chicago Press, pp. 320–339.

Kybourg, Henry E. (1965): "Comment", *Philosophy of Science*, vol. 32, 147–151.

Lange, Marc (2013): "What Makes a Scientific Explanation Distinctively Mathematical?", *British Journal for Philosophy of Science*, vol. 64, pp. 485–511.

Lange, Marc (2014): "Aspects of Mathematical Explanation: Symmetry, Unity, and Salience", *Philosophical Review*, vol. 123, pp. 485–531.

Larkin, Jill H. and Herbert Simon (1987): "Why a Diagram is (Sometimes) Worth Ten Thousand Words", *Cognitive Science*, vol. 11, pp. 65–99.

Laudan, Larry (1977): *Progress and Its Problems. Towards a Theory of Scientific Growth*, Berkeley, Los Angeles and London: University of California Press.

Laudan, Larry (1984): *Science and Values*, Berkeley: University of California Press.

Laudan, Larry (2004): "The Epistemic, the Cognitive, and the Social", in: Peter, Machamer and Gereon, Wolters (eds.): *Science, Values, and Objectivity*, Pittsburgh: University of Pittsburgh Press, pp. 14–23.

Lawson, Tony (1997): *Economics and Reality*, London: Routledge.

Lawson, Tony (2003): *Reorienting Economics*, London: Routledge.

Lewis, David (1986): "Causal Explanation", *Philosophical Papers*, vol. 2, Oxford: Oxford University Press, pp. 214–240.

Lipton, Peter (2004): *Inference to the Best Explanation*, 2nd edition, London and New York: Routledge.

Little, Daniel (2009): "The Heterogeneous Social: New Thinking About the Foundations of the Social Sciences", in C. Mantzavinos (ed.): *Philosophy of* the Social Sciences. Philosophical Theory and Scientific Practice, Cambridge: Cambridge University Press, pp. 154–178.

Lloyd, G.E.R. (2008): "Galen and His Contemporaries", in Robert J. Hankinson (ed.): The Cambridge Companion to Galen, Cambridge: Cambridge University Press, pp. 34–48.

Longino, Helen (1990): Science as Social Knowledge, Princeton: Princeton University Press.

Longino, Helen (2002): The Fate of Knowledge, Princeton: Princeton University Press.

Longino, Helen (2013): Studying Human Behavior: How Scientists Investigate Aggression and Sexuality, Chicago: The University of Chicago Press.

Lorenz, Konrad (1941): "Kants Lehre vom Apriorischen im Lichte gegenwärtiger Biologie", Blätter für deutsche Philosophie, vol. 15 (1), pp. 95–124.

Lorenz, Konrad (1943): "Die angeborenen Formen möglicher Erfahrung", Zeitschrift für Tierpsychologie, vol. 5, pp. 235–409.

Lorenz, Konrad (1965): Evolution and Modification of Behavior, Chicago: Chicago University Press.

Lorenz, Konrad (1973): Die Rückseite des Spiegels. Versuch einer Naturgeschichte menschlicher Erkennens, München and Zürich: Pieper.

Love, Alan C. (2015): "Collaborative Explanation, Explanatory Roles, and Scientific Explaining in Practice", Studies in History and Philosophy of Science, vol. 52, pp. 88–94.

Lütge, Christoph (2001): Ökonomische Wissenschaftstheorie, Würzburg: Königshausen & Neumann.

Lütge, Christoph (2004): "Economics in Philosophy of Science: A Dismal Contribution?" Synthese, vol. 140 (3), pp. 279–305.

Machamer, Peter (2009): "Explaining Mechanisms", unpublished paper.

Machamer, Peter, Lindley Darden, and Carl Craver (2000): "Thinking about Mechanisms", Philosophy of Science, vol. 67, pp. 1–25.

Mäki, Uskali (2001): "Explanatory Unification: Double and Doubtful", Philosophy of the Social Sciences, vol. 31, pp. 488–506.

Mäki, Uskali (2002): "Symposium on Explanations and Social Ontology 2: Explanatory Ecumenism and Economics Imperialism", Economics and Philosophy, vol. 18, pp. 235–257.

Mäki, Uskali and Caterina Marchionni (2009): "On the Structure of Explanatory Unification: The Case of Geographical Economics", Studies in History and Philosophy of Science, vol. 40, pp. 185–195.

Malpighi, Marcello (1661/1929): De Pulmonibus epistolae II ad Borellium, translated into English by James Young: Malpighi's "De Pulmonibus", in Proceedings of the Royal Society of Medicine, vol. 23, pp. 1–11.

Malthus, Thomas Robert (1814): *Observations on the Effects of Corn Laws on the Agriculture and General Wealth of the Country*, London: J. Johnson & Co.

Mantzavinos, C. (1994): *Wettbewerbstheorie. Eine kritische Auseinandersetzung*, Berlin: Duncker & Humblot.

Mantzavinos, C. (1999): "Carl Menger und die Wettbewerbstheorie", *Jahrbücher für Nationalökonomie und Statistik*, vol. 219, pp. 685–691.

Mantzavinos, C. (2001): *Individuals, Institutions and Markets*, Cambridge: Cambridge University Press.

Mantzavinos, C. (2005): *Naturalistic Hermeneutics*, Cambridge: Cambridge University Press.

Mantzavinos, C. (2007): "Interpreting the Rules of the Game", in Christoph Engel and Fritz Strack (eds.): *The Impact of Court Procedure on the Psychology of Judicial Decision Making*, Baden-Baden: Nomos, pp. 13–30.

Mantzavinos, C. (ed.) (2009): *Philosophy of the Social Sciences. Philosophical Theory and Scientific Practice*, Cambridge: Cambridge University Press.

Mantzavinos, C., North Douglass, and Syed Shariq (2004): "Learning, Institutions and Economic Performance", *Perspectives on Politics*, vol. 2, pp. 75–84.

Marchionni, Caterina (2008): "Explanatory Pluralism and Complementarity. From Autonomy to Integration", *Philosophy of the Social Sciences*, vol. 38, pp. 314–333.

Marshall, Alfred (1890/1920): *Principles of Economics*, 8th edition, London: MacMillan.

Mattern, Susan P. (2008): *Galen and the Rhetoric of Healing*, Baltimore: The John Hopkins University Press.

Mayer, Richard E. (1992): *Thinking, Problem Solving, Cognition*, 2nd edition, New York: W.H. Freeman.

McAdams, Richard (1997): "The Origin, Development, and Regulation of Norms", *Michigan Law Review*, vol. 96, pp. 338–433.

McCauley, Robert (2000): "The Naturalness of Religion and the Unnaturalness of Science", in Frank C. Keil and Robert A. Wilson (eds.): *Explanation and Cognition*, Cambridge, MA: The MIT Press, pp. 61–85.

McCauley, Robert (2009): "Time is of the Essence: Explanatory Pluralism and Accommodating Theories About Long Term Processes", *Philosophical Psychology*, vol. 22, pp. 611–635.

McCauley, Robert (2013): "Explanatory Pluralism and the Cognitive Science of Religion: Or Why Scholars in Religious Studies Should Stop Worrying About Reductive Elimination of the Religious and Why Cognitive Science Poses No Threat to Religious Studies", *Working Paper*.

McCauley, Robert and William Bechtel (2001): "Explanatory Pluralism and Heuristic Identity Theory", *Theory and Psychology*, vol. 11, pp. 736–760.

McMullen, Emerson Thomas (1995): "Anatomy of a Physiological Discovery: William Harvey and the Circulation of the Blood", *Journal of the Royal Society of Medicine*, vol. 88, pp. 491–498.

McMullin, Ernan (1983/2012): "Values in Science", in *PSA: Proceedings of the Biennial Meeting of the Philosophy of Science Association*, vol. 182, vol. Two: Symposia and Invited Papers, and reprinted in *Zygon*, vol. 47, pp. 686–709.

McMullin, Ernan (2008): "The Virtues of a Good Theory", in Stathis Psillos and Martin Curd (eds.): *The Routledge Companion to Philosophy of Science*, London and New York: Routledge, pp. 498–508.

Menger, Carl (1871/1968): *Grundsätze der Volkswirtschaftslehre*, Wien: Wilhelm Braumüller, and 2nd edition 1968 by Friedrich A. von Hayek: Carl Menger: Gesammelte Werke, Band I, Tübingen: J.C.B. Mohr (Paul Siebeck).

Menger, Carl (1976): *Principles of Economics*, translated by James Dingwall and Bert F. Hoselitz, Arlington, VA: The Institue for Humane Studies.

Meyerhof, Max (1935): "Ibn An-Nafis (XIIIth cent.) and His Theory of the Lesser Circulation", *Isis*, vol. 23, pp. 100–120.

Mill, John Stuart (1843/1974): *A System of Logic. Ratiocinative and Inductive*, vol. VII of the Collected Works of John Stuart Mill, ed., J.M. Robson and R.F. McRae, Toronto: University of Toronto Press and Routledge & Kegan Paul.

Mill, John Stuart (1848/1909): *Principles of Political Economy with Some of Their Applications to Social Philosophy*, 7th edition, London: Longmans Green and Co.

Mitchell, Sandra (2003): *Biological Complexity and Integrative Pluralism*, Cambridge: Cambridge University Press.

Mitchell, Sandra (2009a): *Unsimple Truths. Science, Complexity and Policy*, Chicago: The University of Chicago Press.

Mitchell, Sandra (2009b): "Complexity and Explanation in the Social Sciences", in: C. Mantzavinos (ed.): *Philosophy of the Social Sciences. Philosophical Theory and Scientific Practice*, Cambridge: Cambridge University Press, pp. 130–145.

Mithen, Steven (1996): *The Prehistory of the Mind: The Cognitive Origins of Art, Religion and Science*, New York: Thames and Hudson.

Mithen, Steven (2002): "Human Evolution and the Cognitive Basis of Sciences", in Peter Carruthers, Stich Stephen, and Michael Siegal (eds.): *The Cognitive Basis of Science*, Cambridge: Cambridge University Press, pp. 23–40.

Morrison, Margaret (2000): *Unifying Scientific Theories*, Cambridge: Cambridge University Press.

Mueller, Dennis (2003): *Public Choice III*, Cambridge: Cambridge University Press.

Nagel, Ernst (1961): *The Structure of Science: Problems in the Logic of Scientific Explanation*, London: Routledge & Kegan Paul.

Nathan, Marco J. (2015): "Unificatory Explanation", *British Journal for the Philosophy of Science* (Advance Access), published July 8, 2015.

Nee, Victor (1998): "Norms and Networks in Economic and Organizational Performance", *American Economic Review (Papers and Proceedings)*, vol. 88, pp. 85–89.

Nee, Victor and Paul Ingram (1998): "Embeddedness and Beyond: Institutions, Exchange and Social Structure", in Victor Nee and Mary C. Brinton (eds.): *The New Institutionalism in Sociology*, New York: Russell Sage Foundation, pp. 19–45.

Nee, Victor and David Strang (1998): "The Emergence and Diffusion of Institutional Forms", *Journal of Institutional and Theoretical Economics*, vol. 154, pp. 706–715.

Nelson, Richard and Sidney Winter (1982): *An Evolutionary Theory of Economic Change*, Cambridge, MA: Harvard University Press.

Nersessian, Nancy (1992): "How Do Scientists Think? Capturing the Dynamics of Conceptual Change in Science", in Ronald Giere (ed.): *Cognitive Models of Science*, Minnesota Studies in the Philosophy of Science, vol. XV, Minneapolis: University of Minnesota Press, pp. 3–44.

Nersessian, Nancy (2008a): *Creating Scientific Concepts*, Cambridge, MA: The MIT Press.

Nersessian, Nancy (2008b): "Mental Modeling in Conceptual Change", in Stella Vosniadou (ed.): *International Handbook of Research on Conceptual Change*, London and New York: Routledge, pp. 391–416.

Neumann, John von and Oskar Morgenstern (1944): *Theory of Games and Economic Behavior*, Princeton: Princeton University Press.

Newell, Allen and Herbert Simon (1972): *Human Problem Solving*, Englewood Cliffs, NJ: Prentice Hall.

Nickel, Bernhard (2010): "How General Do Theories of Explanation Need to Be?" *Noûs*, vol. 42, pp. 305–328.

Nickles, Thomas (2006): "Heuristic Appraisal: Context of Discovery or Justification?", in Jutta Schickore and Friedrich Steinle (eds.): *Revisiting Discovery and Justification*, Dordrecht: Springer, pp. 159–182.

Nietzsche, Friedrich (1887): *Zur Genealogie der Moral. Eine Streitschrift*, Leigzig: Verlag C.G. Naumann.

Nisbett, Richard and Lee Ross (1980): *Human Inference: Strategies and Shortcomings of Social Judgment*, Englewood Cliffs, NJ: Prentice Hall.

North, Douglass C. (1990): *Institutions, Institutional Change, and Economic Performance*, Cambridge: Cambridge University Press.

Nozick, Robert (1994): "Invisible-Hand Explanations", *American Economic Review (Papers and Proceedings)*, vol. 84, pp. 314–318.

Nutton, Vivian (2004): *Ancient Medicine*, London: Routledge.

Nutton, Vivian (2008): "The Fortunes of Galen", in Robert J. Hankinson (ed.): *The Cambridge Companion to Galen*, Cambridge: Cambridge University Press, pp. 355–390.

Olson, Mancur (1965): *The Logic of Collective Action*, Cambridge, MA: Harvard University Press.

O'Malley, Charles Donald (1970): "The Lure of Padua", *Medical History*, vol. 14, pp.1–9.

Opp, Karl-Dieter (1979): "The Emergence and Effects of Social Norms. A Confrontation of Some Hypotheses of Sociology and Economics", *Kyklos*, vol. 32, pp. 775–801.

Opp, Karl-Dieter (1982): "The Evolutionary Emergence of Norms", *British Journal of Social Psychology*, vol. 21, pp. 139–149.

Ott, Alfred E. and Harald Winkel (1985): *Geschichte der theoretischen Volkswirtschaftslehre*, Göttingen: Vandenhoeck & Ruprecht.

Overton, James A. (2013): "'Explain' in Scientific Discourse", *Synthese*, vol. 190, pp. 1383–1405.

Palmieri, Paolo (2012): "Signals, Cochlear Mechanics and Pragmatism: A New Vista on Human Hearing?", *Journal of Experimental & Theoretical Artificial Intelligence*, vol. 24, pp. 527–548.

Papineau, David (2009): "Physicalism and the Human Sciences", in C. Mantzavinos (ed.): *Philosophy of the Social Sciences. Philosophical Theory and Scientific Practice*, Cambridge: Cambridge University Press, pp. 103–123.

Perini, Laura (2005a): "The Truth in Pictures", *Philosophy of Science*, vol. 72, pp. 262–285.

Perini, Laura (2005b): "Visual Representations and Confirmation", *Philosophy of Science*, vol. 72, pp. 913–926.

Perini, Laura (2005c): "Explanation in Two Dimensions: Diagrams and Biological Explanation", *Biology and Philosophy*, vol. 20, pp. 257–269.

Perini, Laura (2013): "Diagrams in Biology", *The Knowledge Engineering Review*, vol. 28, pp. 273–286.

Petkov, Stefan (2015): "Explanatory Unification and Conceptualization", *Synthese*, vol. 192, pp. 3695–3717.

Piaget, Jean (1936): *La Naissance de l' Intelligence chez l'Enfant*, Neuchatel and Paris: Delachau et Niestle.

Pincock, Christopher (2012): *Mathematics and Scientific Representation*, Oxford: Oxford University Press.

Pincock, Christopher (2015): "Abstract Explanations in Science", *British Journal for the Philosophy of Science*, vol. 66, pp. 857–882.

Plutynski, Anya (2004): "Explanation in Classical Population Genetics", *Philosophy of Science*, vol. 71, pp. 1201–1214.

Polanyi, Michael (1958): *Personal Knowledge*, London: Routledge.

Popper Karl R. (1934): *Logik der Forschung*, Vienna: Springer, Imprint 1935, actually published 1934.

Popper, Karl R. (1949): "Naturgesetze und theoretische Systeme," in Simon Moser (ed.): *Gesetze und Wirklichkeit*, Innsbruck: Tyrolia-Verlag, pp. 43–60.

Popper, Karl R. (1957): "The Aim of Science", *Ratio*, I, pp. 24–35 [and reprinted in *Objective Knowledge. An Evolutionary Approach*, rev. edition, Oxford: Oxford University Press, 1979, pp. 191–205].

Popper, Karl R. (1958/1989): "Back to the Presocratics", in *Proceedings of the Aristotelian Society*, N.S. vol. 59, and reprinted in: *Conjectures and Refutations. The Growth of Scientific Knowledge*, 5th revised edition, London: Routledge, pp. 136–165.

Popper, Karl R. (1959/2002): *The Logic of Scientific Discovery*, London: Routledge.

Popper, Karl R. (1963/1989): *Conjectures and Refutations. The Growth of Scientific Knowledge*, 5th revised edition, London: Routledge.

Popper, Karl R. (1970): "Normal Science and Its Dangers", in Imre Lakatos and Alan Musgrave (eds.): *Criticism and the Growth of Knowledge*, Cambridge: Cambridge University Press, pp. 51–58.

Popper, Karl R. (1972/1992): *Objective Knowledge. An Evolutionary Approach*, Oxford: Clarendon Press.

Popper, Karl R. (1987): "Die erkenntnistheoretische Position der evolutionären Erkenntnistheorie", in Rupert Riedl and Franz Wuketits (eds.): *Die Evolutionäre Erkenntnistheorie. Bedingungen. Lösungen. Kontroversen*, Berlin: Parcy.

Popper, Karl R. and John C. Eccles (1977/1983): *The Self and Its Brain. An Argument for Interactionism*, London: Routledge.

Posner, Richard (2010): *Economic Analysis of Law*, 8th edition, New York: Aspen Publishers.

Potochnik, Angela (2010): "Levels of Explanation Reconceived", *Philosophy of Science*, vol. 77, pp. 59–72.

Prioreschi, Plinio (2004): "Andrea Cesalpino and Systemic Circulation", *Annales Pharmaceutiques Francaises*, vol. 62, pp. 382–400.

Psillos, Stathis (2002): *Causation and Explanation*, Montreal and Kingston: McGill-Queen's University Press.

Psillos, Stathis (2007): "Past and Contemporary Perspectives on Explanation", in Theo Kuipers (ed.): *Handbook of the Philosophy of Science. Focal Issues*, Amsterdam and Oxford: Elsevier, pp. 97–174.

Railton, Peter (1978): "A Deductive-Nomological Model of Probabilistic Explanation", *Philosophy of Science*, vol. 45, pp. 206–226.

Railton, Peter (1981): "Probability, Explanation and Information", *Synthese*, vol. 48, pp. 233–256.

Reutlinger, Alexander (2014): "Why Is There Universal Macrobehavior? Renormalization Group Explanation as Noncausal Explanation", *Philosophy of Science*, vol. 81, pp. 1157–1170.

Ricardo, David (1817/1951): *On the Principles of Political Economy and Taxation*, ed. Pierro Sraffa, Cambridge: Cambridge University Press.

Rice, Collin (2015): "Moving Beyond Causes: Optimality Models and Scientific Explanation", *Noûs*, vol. 49, pp. 589–615.

Richardson, Alan W. (2006): "The Many Unities of Science: Politics, Semantics, and Ontology", in Stephen H. Kellert, Helen E. Longino, and C. Kenneth Waters (eds.): *Scientific Pluralism*, Minneapolis and London: University of Minnesota Press, pp. 1–25.

Rocca, Julius (2008): "Anatomy", in Robert J. Hankinson (ed.): *The Cambridge Companion to Galen*, Cambridge: Cambridge University Press, pp. 242–262.

Rogers, Brian and Cassandra, Rogers (2009): "Visual Globes, Celestial Spheres, and the Perception of Straight and Parallel Lines", *Perception*, vol. 38, pp. 1295–1312.

Rosefeld, Tobias (2004): "Is Knowing-how Simply a Case of Knowing-that?" *Philosophical Investigations*, vol. 27, pp. 370–379.

Roth, Moritz (1892): *Andreas Vesalius Bruxellensis*, Berlin: George Reimer.

Ruben, David-Hillel (2012): *Explaining Explanation*, 2nd edition, Boulder and London: Paradigm Publishers.

Ruphy, Stéphanie (2011): "From Hacking's Plurality of Styles of Scientific Reasoning to 'Foliated Pluralism': A Philosophically Robust Form of Ontologico-Methodological Pluralism", *Philosophy of Science*, vol. 78, pp. 1212–1222.

Russell, Bertrand (1912): "On the Notion of Cause", *Proceedings of the Aristotelian Society New Series*, vol. 13, pp. 1–26.

Ryle, Gilbert (1949): *The Concept of Mind*, London: Penguin Books.

Saatsi, Juha and Mark Pexton (2013): "Reassessing Woodward's Account of Explanation: Regularities, Counterfactuals, and Noncausal Explanations", *Philosophy of Science*, vol. 80, pp. 613–624.

Salmon, Wesley (1970): "Statistical Explanation", in Robert G. Colodny (ed.): *The Nature and Function of Scientific Theories*, Pittsburgh: University of

Pittsburgh Press, pp. 173–231. [Reprinted in Wesley Salmon (1971): *Statistical Explanation and Statistical Relevance*, Pittsburgh: University of Pittsburgh Press, pp. 29–87.]

Salmon, Wesley (1984): *Scientific Explanation and the Causal Structure of the World*, Princeton: Princeton University Press.

Salmon, Wesley (1990): *Four Decades of Scientific Explanation*, Minneapolis: University of Minnesota Press.

Salmon, Wesley (1998): *Causality and Explanation*, Oxford: Oxford University Press.

Samarapungavan, Ala (1992): "Children's Judgments in Theory Choice Tasks: Scientific Rationality in Childhood", *Cognition*, vol. 45, pp. 1–32.

Schank, Roger and Robert Abelson (1977): *Scripts, Plans, Goals and Understanding*, Hillsdale, NJ: Erlbaum.

Schiffer, Stephen (2002): "Amazing Knowledge", *Journal of Philosophy*, vol. 99, pp. 200–202.

Schmid, Michael (2006): *Die Logik mechanismischer Erklärungen*, Wiesbaden: VS Verlag für Sozialwissenschaften.

Schrödinger, Erwin (1958): *Mind and Matter*, Cambridge: Cambridge University Press.

Schupbach, Jonah N. and Jan, Sprenger (2011): "The Logic of Explanatory Power", *Philosophy of Science*, vol. 78, pp. 105–127.

Schurz, Gerhard (1999): "Explanation as Unification", *Synthese*, vol. 120, pp. 95–114.

Schurz, Gerhard and Karel Lambert (1994): "Outline of a Theory of Scientific Understanding", *Synthese*, vol. 101, pp. 65–120.

Servetus, Miguel (1553): *Christianismi Restitutio*, Vienne, English translation: *The Restoration of Christianity*, by Christopher A. Hoffmann and Marian Hiller, Lewiston, NY: The Edwin Mellen Press, 2008.

Sextus, Empiricus (1933): *Outlines of Pyrrhonism*, translated by R.G. Bury, Loeb Classical Library, Cambridge, MA: Harvard University Press.

Sheredos, Benjamin, Daniel Burnston, Adele Abrahamsen and William Bechtel (2013): "Why Do Biologists Use So Many Diagrams?", *Philosophy of Science*, vol. 80, pp. 931–944.

Shulman, Robert G. and Ian Shapiro (2009): "Reductionism in the Human Sciences: A Philosopher's Game", in C. Mantzavinos (ed.): *Philosophy of the Social Sciences. Philosophical Theory and Scientific Practice*, Cambridge: Cambridge University Press, pp. 124–129.

Simon, Herbert (1992): "Scientific Discovery as Problem Solving", in Massimo Egidi and Robin Marris (eds.): *Economics, Bounded Rationality and the Cognitive Revolution*, Hants: Edward Elgar, pp. 102–119.

Simon, Herbert (2000): "Discovering Explanations", in Frank C. Keil and Robert A. Wilson (eds.): *Explanation and Cognition*, Cambridge, MA: The MIT Press, pp. 21–59.

Siraisi, Nancy (1981): *Tadeo Alderotti and His Pupils: Two Generations of Italian Medical Learning*, Princeton: Princeton University Press.

Skow, Bradford (2014): "Are There Non-Causal Explanations (of Particular Events?)" *British Journal for the Philosophy of Science*, vol. 65, pp. 445–467.

Skow, Bradford (2015): "Are There Genuine Physical Explanations of Mathematical Phenomena?" *British Journal for the Philosophy of Science*, vol. 66, pp. 69–93.

Smith, Adam (1759/1976): *The Theory of Moral Sentiments*, Oxford: Oxford University Press.

Smith, Adam (1776/1976): *An Inquiry into the Nature and Causes of the Wealth of Nations*, ed. Edwin Cannan, Chicago: The University of Chicago Press.

Smith, Ryan (2014): "Explanation, Understanding, and Control", *Synthese*, vol. 191, pp. 4169–4200.

Snellen, H.A. (1984): *History of Cardiology*, Rotterdam: Donker Academic Publications.

Sober, Elliott (1983): "Equilibrium Explanation", *Philosophical Studies*, vol. 43, pp. 201–210.

Sober, Elliott (2002): "What is the Problem of Simplicity?" in Arnold Zellner, Hugo Keuzenbach and Michael McAleer (eds.): *Simplicity, Inference and Econometric Modelling*, Cambridge: Cambridge University Press, pp. 13–31.

Sokal, Alan and Jean Bricmont (1997): *Les Impostures Intellectuelles*, Paris: Editions Odile Jacob.

Staden, Heinrich von (1995): "Anatomy as Rhetoric: Galen on Dissection and Persuasion", *Journal of the History of Medicine and Allied Sciences*, vol. 50, pp. 47–66.

Stanley, Jason (2011): *Know How*, Oxford: Oxford University Press.

Stanley, Jason and Timothy Williamson (2001): "Knowing How", *The Journal of Philosophy*, vol. 98, pp. 411–444.

Steel, Daniel (2004): "Can a Reductionist Be a Pluralist?", *Biology and Philosophy*, vol. 19, pp. 55–73.

Steiner, Mark (1978a): "Mathematical Explanation", *Philosophical Studies*, vol. 34, pp. 135–151.

Steiner, Mark (1978b): "Mathematics, Explanation, and Scientific Knowledge", *Noûs*, vol. 12, pp. 17–28.

Sterelny, Kim (1996): "Explanatory Pluralism in Evolutionary Biology", *Biology and Philosophy*, vol. 11, pp. 193–214.

Sterelny, Kim (2014): *The Evolved Apprentice. How Evolution Made Humans Unique*, Cambridge, MA, and London: The MIT Press.

Streissler, Erich (1989): "Carl Menger", in Joachim Starbatty (ed.): *Klassiker des ökonomischen Denkens*, vol. II, München: Beck, pp. 118–127.

Strevens, Michael (2008): *Depth. An Account of Scientific Explanation*, Cambridge, MA: Harvard University Press.

Stump, David (1992): "Naturalized Philosophy of Science with a Plurality of Methods", *Philosophy of Science*, vol. 59, pp. 456–460.

Suárez, Mauricio (2004): "An Inferential Conception of Scientific Representation", *Philosophy of Science*, vol. 71, pp. 767–779.

Suárez, Mauricio (2015): "Deflationary Representation, Inference, and Practice", *Studies in History and Philosophy of Science*, vol. 49, pp. 36–47.

Sugden, Robert (1998): "Normative Expectations: The Simultaneous Evolution of Institutions and Norms", in Avner Ben-Ner and Louis Putterman (eds.): *Economics, Values, and Organization*, Cambridge: Cambridge University Press, pp. 73–100.

Sun, Ron, Edward Merrill, and Todd Peterson (2001): "From Implicit Skills to Explicit Knowledge: A Bottom-Up Model of Skill Learning", *Cognitive Science*, vol. 25, pp. 203–244.

Suppes, Patrick (1978): "The Plurality of Science", *Philosophy of Science, Proceedings of the Biennial Meeting of the Philosophy of Science Association, Vol. 2: Symposia and Invited Papers*, pp. 3–16.

Teller, Paul (2008): "Representation in Science", in Stathis Psillos and Martin Curd (eds.): *The Routledge Companion to Philosophy of Science*, London and New York, pp. 435–441.

Thagard, Paul (1988): *Computational Philosophy of Science*, Cambridge, MA: The MIT Press.

Thagard, Paul (1996): *Mind. An Introduction into Cognitive Science*, Cambridge, MA: The MIT Press.

Thagard, Paul (2012): *The Cognitive Science of Science. Explanation, Discovery and Conceptual Change*, Cambridge, MA: The MIT Press.

Thalos, Mariam (2002): "Explanation is a Genus: An Essay on the Varieties of Scientific Explanation", *Synthese*, vol. 130, pp. 317–354.

Thiene, Gaetano (1997): "The Discovery of Circulation and the Origin of Modern Medicine during the Italian Renaissance", *Cardiovascular Pathology*, vol. 6 (2), March/April, pp. 79–88.

Thünen, Johann Heinrich von (1842): *Der isolierte Staat in Beziehung auf Landwirtschaft und Nationalökonomie, zweite erweiterte Ausgabe*, Rostock: Leopold.

Topitsch, Ernst (1958): *Vom Ursprung und Ende der Metaphysik. Eine Studie zur Weltanschauungskritik*, Wien: Springer Verlag.

Trivers, Robert (1971): "The Evolution of Reciprocal Altruism", *Quarterly Review of Biology*, vol. 46, pp. 45–57.

Ullmann-Margalit, Edna (1978): "Invisible-Hand Explanations", *Synthese*, vol. 39, pp. 263–291.

van Bouwel, Jeroen (2004): "Explanatory Pluralism in Economics: Against the Mainstream?" *Philosophical Explorations*, vol. 7, pp. 299–315.

van Bouwel, Jeroen and Erik Weber (2008): "A Pragmatist Defense of Non-Relativistic Explanatory Pluralism in History and Social Science", *History and Theory*, vol. 47, pp. 168–182.

van Fraassen, Bas C. (1980): *The Scientific Image*, Oxford: Clarendon Press.

van Fraassen, Bas C. (2008): *Scientific Representation*, Oxford: Oxford University Press.

Vesalius, Andreas (1538): *Tabulae Anatomicae Sex*, Venice.

Vesalius, Andreas (1543): *De Humani Corporis Fabrica Libri Septem*, Basel: Johannes Oporinus, English Translation: *On the Fabric of the Human Body*, by William Frank Richardson and John Burd Carman, volume 5 containing *Books VI: The Heart and Associated Organs and Book VI: The Brain*, Novato CA: Norman Anatomy Series, No 5.

Vollmer, Gerhard (1975/2002): *Evolutionäre Erkenntnistheorie*, 8th edition, Stuttgart: Hirzel.

Vosniadou, Stella and William F. Brewer (1992): "Mental Models of the Earth: A Study of Conceptual Change in Childhood", *Cognitive Psychology*, vol. 24, pp. 535–585.

Vosniadou, Stella and William F. Brewer (1994): "Mental Models of the Day/Night Cycle", *Cognitive Science*, vol. 18, pp. 123–183.

Vosniadou, Stella, Xenia Vamvakoussi, and Irini Skopeliti (2008): "The Framework Theory Approach to the Problem of Conceptual Change", in Stella Vosniadou (ed.): *International Handbook of Research on Conceptual Change*, New York and London: Routledge, pp. 3–34.

Wallis, Charles (2008): "Consciousness, Context, and Know-How", *Synthese*, vol. 160, pp. 123–153.

Walras, Leon (1874): *Éléments d'Économie Politique Pure, ou Théorie de la Richesse Sociale*, Lausanne: L. Corbaz & Cie.

Weatherall, James Owen (2011): "On (Some) Explanations in Physics", *Philosophy of Science*, vol. 78, pp. 421–447.

Weber, Erik and Jeroen van Bouwel (2002): "Symposium on Explanations and Social Ontology 3: Can We Dispense with Structural Explanations of Social Facts?", *Economics and Philosophy*, vol. 18, pp. 259–275.

Weber Erik, Jeroen van Bouwel, and Leen de Vreese (2013): *Scientific Explanation*, Heidelberg, New York and London: Springer.

Weber, Max (1985): *Gesammelte Aufsätze für Wissenschaftslehre*, 6th revised edition, Tübingen: J.C.B. Mohr (Paul Siebeck).

Weslake, Brad (2010): "Explanatory Depth", *Philosophy of Science*, vol. 77, pp. 273–294.

Whitehead, Alfred N. (1911): *An Introduction to Mathematics*, London: Williams and Norgate.

Wiberg, Julius (1937): "The Medical Science of Ancient Greece: The Doctrine of the Heart", *Janus*, vol. 41, pp. 225–254.

Wilkenfeld, Daniel A. (2014): "Functional Explaining. A New Approach to the Philosophy of Explanation", *Synthese*, vol. 191, pp. 3367–3391.

Willius, Fredrick A. and Thomas J. Dry (1948): *A History of the Heart and the Circulation*, Philadelphia: WB Saunders Co.

Wimsatt, William C. (1976): "Reductionism, Levels of Organization, and the Mind-Body Problem", in G. Globus, G. Maxwell and I. Savodnik (eds.), *Consciousness and the Brain: A Scientific and Philosophical Inquiry*, New York: Plenum, pp. 202–267.

Wimsatt, William C. (2007): *Re-Engineering Philosophy for Limited Beings. Piecewise Approximations to Reality*, Cambridge, MA: Harvard University Press.

Winter, Susan (2013): *The Prince of Medicine: Galen in the Roman Empire*, Oxford: Oxford University Press.

Wittgenstein, Ludwig (1953): *Philosophische Untersuchungen/Philosophical Investigations*, translated by G.E.M. Anscombe, Oxford: Blackwell.

Wong, Wai-hung and Zanja, Yudell (2015): "A Normative Account of the Need for Explanation", *Synthese*, vol. 192, pp. 2863–2885.

Woodward, James (2000): "Explanation and Invariance in the Special Sciences," *British Journal for the Philosophy of Science*, vol. 51, pp. 197–254.

Woodward, James (2003): *Making Things Happen*, Oxford: Oxford University Press.

Woodward, James (2014): "Interventionism and Causal Exclusion", *Philosophy and Phenomenological Research*, DOI:10.1111/phpr.12095.

Woody, Andrea I. (2004): "More Telltale Signs: What Attention to Representation Reveals About Scientific Explanation", *Philosophy of Science*, vol. 71, pp. 780–793.

Woody, Andrea I. (2015): "Re-orienting Discussions of Scentific Explanation: A Functional Perspective", *Studies in History and Philosophy of Science*, vol. 52, pp. 79–87.

Wright, Cory (2015): "The Ontic Conception of Scientific Explanation", *Studies in History and Philosophy of Science*, vol. 54, pp. 20–30.

Ylikoski, Petri and Jaako Kuorikoski (2010): "Dissecting Explanatory Power", *Philosophical Studies*, vol. 148, pp. 201–219.

Name Index

Subject Index

www.ingramcontent.com/pod-product-compliance
Ingram Content Group UK Ltd.
Pitfield, Milton Keynes, MK11 3LW, UK
UKHW021959190125
453752UK00006B/25